Standards Practice Book
For Home or School
Grade 5

Houghton
Mifflin
Harcourt

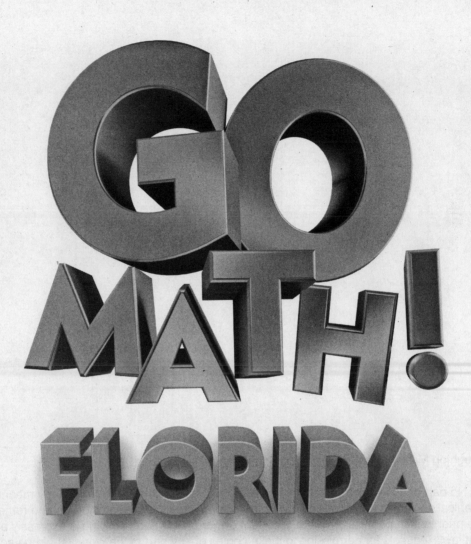

INCLUDES:

- Home or School Practice
- Lesson Practice and Test Preparation
- English and Spanish School-Home Letters
- Getting Ready for Grade 6 Lessons

Fluency with Whole Numbers and Decimals

Extending division to 2-digit divisors, integrating decimal fractions into the place value system and developing understanding of operations with decimals to hundredths, and developing fluency with whole number and decimal operations

1 Place Value, Multiplication, and Expressions

Domains Number and Operations in Base Ten
Operations and Algebraic Thinking

2 Divide Whole Numbers

Domains Number and Operations in Base Ten

5 Divide Decimals

Domain Number and Operations in Base Ten

Operations with Fractions

Developing fluency with addition and subtraction of fractions, and developing understanding of the multiplication of fractions and of division of fractions in limited cases (unit fractions divided by whole numbers and whole numbers divided by unit fractions)

6 Add and Subtract Fractions with Unlike Denominators

Domain Number and Operations–Fractions

© Houghton Mifflin Harcourt Publishing Company

7 Multiply Fractions

Domain Number and Operations–Fractions

8 Divide Fractions

Domain Number and Operations–Fractions

Geometry and Measurement

Developing understanding of volume

9 — Algebra: Patterns and Graphing

Domains Measurement and Data
Geometry
Operations and Algebraic Thinking

10 — Convert Units of Measure

Domain Measurement and Data

11 Geometry and Volume

Domains Measurement and Data
Geometry

End-of-Year Resources

Getting Ready for Grade 6

These lessons review important skills and prepare you for Grade 6.

School-Home Letter

© Houghton Mifflin Harcourt Publishing Company

Dear Family,

Throughout the next few weeks, our math class will be learning about place value, number properties, and numerical expressions. We will also learn to multiply by 1- and 2-digit whole numbers.

You can expect to see homework that requires students to write and evaluate numerical expressions.

Here is a sample of how your child will be taught to evaluate an expression.

Vocabulary

evaluate To find the value of a numerical or algebraic expression

numerical expression A mathematical phrase that has numbers and operation signs but does not have an equal sign

order of operations The process for evaluating expressions

🔑 MODEL Evaluate Expressions

This is how we will be evaluating $36 - (2 + 3) \times 4$.

STEP 1	
Perform the operations in parentheses.	$36 - (2 + 3) \times 4$ $36 - 5 \times 4$
STEP 2	
Multiply.	$36 - 20$
STEP 3	
Subtract.	**16**

$36 - (2 + 3) \times 4 = 16$

Tips

Order of Operations

To evaluate an expression, first perform the operations in parentheses. Next, multiply and divide from left to right. Finally, add and subtract from left to right.

Activity

You can write numerical expressions to describe situations around the house. For example, "We bought a case of 24 water bottles and have used 13 bottles. What expression shows how many are left?" can be represented by the expression $24 - 13$.

Carta para la casa

Vocabulario

evaluar Hallar el valor de una expresión numérica o algebraica

expresión numérica Una frase matemática que tiene solo números y signos de operaciones.

orden de las operaciones El proceso que se usa para evaluar expresiones

Querida familia,

Durante las próximas semanas, en la clase de matemáticas aprenderemos sobre el valor de posición, las propiedades de los números y las expresiones numéricas.

Llevaré a la casa tareas con actividades para practicar la escritura y evaluación de expresiones numéricas.

Este es un ejemplo de la manera en que evaluaremos expresiones numéricas.

🔓 MODELO Evaluar expresiones

Así es como evaluaremos $36 - (2 + 3) \times 4$.

Sandra tiene 8 manzanas. Le da algunas manzanas a Josh.

PASO 1

Resuelve las operaciones en paréntesis.

$36 - (2 + 3) \times 4$
$36 - 5 \times 4$

PASO 2

Multiplica.

$36 - 20$

PASO 3

Resta.

16

$36 - (2 + 3) \times 4 = 16$

Pistas

Orden de las Operaciones

Para evaluar una expresión, primero resuelve las operaciones en paréntesis. Después multiplica y divide de izquierda a derecha. Finalmente suma y resta de izquierda a derecha.

Actividad

Pueden escribir expresiones numéricas para representar cosas que suceden en la casa. Por ejemplo, "Compramos una caja de 24 botellas de agua y usamos 13 botellas. ¿Qué expresión muestra cuántas botellas quedan?", se puede representar con $24 - 13$.

Name _____

Place Value and Patterns

Complete the sentence.

1. 40,000 is 10 times as much as ___4,000___.

2. 90 is $\frac{1}{10}$ of _____.

3. 800 is 10 times as much as _____.

4. 5,000 is $\frac{1}{10}$ of _____.

Use place-value patterns to complete the table.

Number	10 times as much as	$\frac{1}{10}$ of
5. 100		
6. 7,000		
7. 300		
8. 80		

Number	10 times as much as	$\frac{1}{10}$ of
9. 2,000		
10. 900		
11. 60,000		
12. 500		

Problem Solving  REAL WORLD

13. The Eatery Restaurant has 200 tables. On a recent evening, there were reservations for $\frac{1}{10}$ of the tables. How many tables were reserved?

14. Mr. Wilson has $3,000 in his bank account. Ms. Nelson has 10 times as much money in her bank account as Mr. Wilson has in his bank account. How much money does Ms. Nelson have in her bank account?

_____   _____

Lesson Check

1. What is 10 times as much as 700?

   (A)   7

   (B)   70

   (C)   7,000

   (D) 70,000

2. What is $\frac{1}{10}$ of 3,000?

   (A) 30,000

   (B)   300

   (C)   30

   (D)   3

Spiral Review

3. Risa is sewing a ribbon around the sides of a square blanket. Each side of the blanket is 72 inches long. How many inches of ribbon will Risa need? (Grade 4)

   (A)   144 inches

   (B)   208 inches

   (C)   288 inches

   (D) 5,184 inches

4. What is the value of n? (Grade 4)

   $$9 \times 27 + 2 \times 31 - 28 = n$$

   (A)   249

   (B)   277

   (C)   783

   (D) 7,567

5. Between what pair of numbers is the product of 289 and 7? (Grade 4)

   (A) between 200 and 300

   (B) between 1,400 and 1,500

   (C) between 1,400 and 1,800

   (D) between 1,400 and 2,100

6. Which list shows the numbers in order from **greatest** to **least**? (Grade 4)

   (A) 7,613; 7,361; 7,136

   (B) 7,631; 7,136; 7,613

   (C) 7,136; 7,361; 7,613

   (D) 7,136; 7,613; 7,361

Name _____

Place Value of Whole Numbers

Write the value of the underlined digit.

1. 5,1<u>6</u>5,874

__60,000__

2. 2<u>8</u>1,480,100

_____

3. 7,<u>2</u>70

_____

4. 89,<u>1</u>70,326

_____

5. <u>7</u>,050,423

_____

6. 6<u>4</u>6,950

_____

7. 37,<u>1</u>23,745

_____

8. <u>3</u>15,421,732

_____

Write the number in two other forms.

9. 15,409

_____

_____

_____

10. 100,203

_____

_____

_____

11. 6,007,200

_____

_____

_____

12. 32,005,008

_____

_____

_____

Problem Solving  REAL WORLD

13. The U.S. Census Bureau has a population clock on the Internet. On a recent day, the United States population was listed as 310,763,136. Write this number in word form.

_____

_____

_____

14. In 2008, the population of 10- to 14-year-olds in the United States was 20,484,163. Write this number in expanded form.

_____

_____

_____

Lesson Check

1. A movie cost $3,254,107 to produce. Which digit is in the hundred thousands place?

   Ⓐ 5

   Ⓑ 4

   Ⓒ 2

   Ⓓ 1

2. Which is another way to write two hundred ten million, sixty-four thousand, fifty?

   Ⓐ 210,640,050

   Ⓑ 210,064,050

   Ⓒ 201,064,500

   Ⓓ 200,106,450

Spiral Review

3. If the pattern below continues, what number likely comes next? **(Grade 4)**

   9, 12, 15, 18, 21, ?

   Ⓐ 36

   Ⓑ 24

   Ⓒ 22

   Ⓓ 20

4. What is 52 ÷ 8? **(Grade 4)**

   Ⓐ 8 r4

   Ⓑ 7 r4

   Ⓒ 6 r4

   Ⓓ 5 r4

5. How many pairs of parallel sides does the trapezoid below have? **(Grade 4)**

   Ⓐ 0

   Ⓑ 1

   Ⓒ 2

   Ⓓ 4

6. Which figure appears to have only 1 line of symmetry? **(Grade 4)**

   Ⓐ

   Ⓑ

   Ⓒ

   Ⓓ

Name _____

Properties

Use properties to find the sum or product.

1. 6×89

$6 \times (90 - 1)$

$(6 \times 90) - (6 \times 1)$

$540 - 6$

534

2. $93 + (68 + 7)$

_____

3. $5 \times 23 \times 2$

_____

4. 8×51

_____

5. $34 + 0 + 18 + 26$

_____

6. 6×107

_____

Complete the equation, and tell which property you used.

7. $(3 \times 10) \times 8 = $ _____ $\times (10 \times 8)$

_____

_____

8. $16 + 31 = 31 + $ _____

_____

_____

9. $0 + $ _____ $ = 91$

_____

_____

10. $21 \times $ _____ $ = 9 \times 21$

_____

_____

Problem Solving REAL WORLD

11. The Metro Theater has 20 rows of seats with 18 seats in each row. Tickets cost $5. The theater's income in dollars if all seats are sold is $(20 \times 18) \times 5$. Use properties to find the total income.

_____

12. The numbers of students in the four sixth-grade classes at Northside School are 26, 19, 34, and 21. Use properties to find the total number of students in the four classes.

_____

Lesson Check

1. To find 19 + (11 + 37), Lennie added 19 and 11. Then he added 37 to the sum. Which property did he use?

   Ⓐ Distributive Property

   Ⓑ Commutative Property of Addition

   Ⓒ Associative Property of Addition

   Ⓓ Identity Property of Addition

2. Marla did 65 sit-ups each day for one week. Which expression can you use to find the total number of sit-ups Marla did during the week?

   Ⓐ $(7 \times 6) + (7 \times 5)$

   Ⓑ $(5 \times 60) + (5 \times 7)$

   Ⓒ $(7 + 60) \times (7 + 5)$

   Ⓓ $(7 \times 60) + (7 \times 5)$

Spiral Review

3. The average sunflower has 34 petals. Which is the best estimate of the total number of petals on 57 sunflowers? **(Grade 4)**

   Ⓐ 18

   Ⓑ 180

   Ⓒ 1,800

   Ⓓ 18,000

4. A golden eagle flies a distance of 290 miles in 5 days. If the eagle flies the same distance each day of its journey, how far does the eagle fly per day? **(Grade 4)**

   Ⓐ 50 miles

   Ⓑ 58 miles

   Ⓒ 290 miles

   Ⓓ 295 miles

5. What is the value of the underlined digit in the following number? **(Lesson 1.2)**

   2,9<u>8</u>3,785

   Ⓐ 80

   Ⓑ 800

   Ⓒ 8,000

   Ⓓ 80,000

6. The number 5 is **(Grade 4)**

   Ⓐ prime.

   Ⓑ composite.

   Ⓒ neither prime nor composite.

   Ⓓ both prime and composite.

Name _____

Powers of 10 and Exponents

Write in exponent form and word form.

1. $10 \times 10 \times 10$

2. 10×10

3. $10 \times 10 \times 10 \times 10$

exponent form: ___10^3___

word form: __the third power of ten__

exponent form: _____

word form: _____

exponent form: _____

word form: _____

Find the value.

4. 10^3

5. 4×10^2

6. 9×10^4

7. 10^1

_____

8. 10^5

9. 5×10^1

10. 7×10^3

11. 8×10^0

_____

Problem Solving  REAL WORLD

12. The moon is about 240,000 miles from Earth. What is this distance written as a whole number multiplied by a power of ten?

_____

13. The sun is about 93×10^6 miles from Earth. What is this distance written as a whole number?

_____

Lesson Check

1. Which of the following is NOT equivalent to "3 times the sixth power of 10?"

   (A) 3×10^6

   (B) 3,000,000

   (C) $3 \times 10 \times 6$

   (D) $3 \times 1,000,000$

2. Gary mails 10^3 flyers to clients in one week. How many flyers does Gary mail?

   (A) 10

   (B) 100

   (C) 1,000

   (D) 10,000

Spiral Review

3. Harley is loading 625 bags of cement onto small pallets. Each pallet holds 5 bags. How many pallets will Harley need? **(Grade 4)**

   (A) 125

   (B) 620

   (C) 630

   (D) 3,125

4. Marylou buys a package of 500 jewels to decorate 4 different pairs of jeans. She uses the same number of jewels on each pair of jeans. How many jewels will she use for each pair of jeans? **(Grade 4)**

   (A) 100

   (B) 125

   (C) 200

   (D) 2,000

5. Manny buys 4 boxes of straws for his restaurant. There are 500 straws in each box. How many straws does he buy? **(Grade 4)**

   (A) 20,000

   (B) 2,000

   (C)   200

   (D) 125

6. Cammie goes to the gym to exercise 4 times per week. Altogether, how many times does she go to the gym in 10 weeks? **(Grade 4)**

   (A) 4

   (B) 10

   (C) 20

   (D) 40

Name _____

Multiplication Patterns

Use mental math to complete the pattern.

1. $8 \times 3 = 24$

   $(8 \times 3) \times 10^1 =$ **240**

   $(8 \times 3) \times 10^2 =$ **2,400**

   $(8 \times 3) \times 10^3 =$ **24,000**

2. $5 \times 6 =$ _____

   $(5 \times 6) \times 10^1 =$ _____

   $(5 \times 6) \times 10^2 =$ _____

   $(5 \times 6) \times 10^3 =$ _____

3. $3 \times$ _____ $= 27$

   $(3 \times 9) \times 10^1 =$ _____

   $(3 \times 9) \times 10^2 =$ _____

   $(3 \times 9) \times 10^3 =$ _____

4. _____ $\times 4 = 28$

   $(7 \times 4) \times$ _____ $= 280$

   $(7 \times 4) \times$ _____ $= 2,800$

   $(7 \times 4) \times$ _____ $= 28,000$

5. $6 \times 8 =$ _____

   $(6 \times 8) \times 10^2 =$ _____

   $(6 \times 8) \times 10^3 =$ _____

   $(6 \times 8) \times 10^4 =$ _____

6. _____ $\times 4 = 16$

   $(4 \times 4) \times 10^2 =$ _____

   $(4 \times 4) \times 10^3 =$ _____

   $(4 \times 4) \times 10^4 =$ _____

Use mental math and a pattern to find the product.

7. $(2 \times 9) \times 10^2 =$ _____

8. $(8 \times 7) \times 10^2 =$ _____

9. $(9 \times 6) \times 10^3 =$ _____

10. $(3 \times 7) \times 10^3 =$ _____

11. $(5 \times 9) \times 10^4 =$ _____

12. $(4 \times 8) \times 10^4 =$ _____

13. $(8 \times 8) \times 10^3 =$ _____

14. $(6 \times 4) \times 10^4 =$ _____

15. $(5 \times 5) \times 10^3 =$ _____

Problem Solving REAL WORLD

16. The Florida Everglades welcomes about 2×10^3 visitors per day. Based on this, about how many visitors come to the Everglades per week?

    _____

    _____

17. The average person loses about 8×10^1 strands of hair each day. About how many strands of hair would the average person lose in 9 days?

    _____

    _____

Lesson Check

1. How many zeros are in the product $(6 \times 5) \times 10^3$?

   Ⓐ 3
   Ⓑ 4
   Ⓒ 5
   Ⓓ 6

2. Addison studies a tarantula that is 30 millimeters long. Suppose she uses a microscope to magnify the spider by 4×10^2. How long will the spider appear to be?

   Ⓐ   12 millimeters
   Ⓑ   120 millimeters
   Ⓒ   1,200 millimeters
   Ⓓ   12,000 millimeters

Spiral Review

3. Hayden has 6 rolls of dimes. There are 50 dimes in each roll. How many dimes does he have altogether? **(Grade 4)**

   Ⓐ 300
   Ⓑ 110
   Ⓒ   56
   Ⓓ   30

4. An adult ticket to the zoo costs $20, and a child's ticket costs $10. How much will it cost for Mr. and Mrs. Brown and their 4 children to get into the zoo?

   (Grade 4)

   Ⓐ $40
   Ⓑ $60
   Ⓒ $80
   Ⓓ $100

5. At a museum, 100 posters are displayed in each of 4 rooms. Altogether, how many posters are displayed? **(Grade 4)**

   Ⓐ   40
   Ⓑ 100
   Ⓒ 104
   Ⓓ 400

6. A store sells a gallon of milk for $3. A baker buys 30 gallons of milk for his bakery. How much will he have to pay? **(Grade 4)**

   Ⓐ $120
   Ⓑ $90
   Ⓒ $60
   Ⓓ $30

Name _____

Multiply by 1-Digit Numbers

Estimate. Then find the product.

1. Estimate: __3,600__

$$
\begin{array}{r}
{}^{1\,5}\\
416\\
\times\ \ 9\\
\hline
3{,}744
\end{array}
$$

2. Estimate: _____

$$
\begin{array}{r}
1{,}374\\
\times\ \ 6\\
\hline
\end{array}
$$

3. Estimate: _____

$$
\begin{array}{r}
726\\
\times\ \ 5\\
\hline
\end{array}
$$

4. Estimate: _____

$$
\begin{array}{r}
872\\
\times\ \ 3\\
\hline
\end{array}
$$

5. Estimate: _____

$$
\begin{array}{r}
2{,}308\\
\times\ \ 9\\
\hline
\end{array}
$$

6. Estimate: _____

$$
\begin{array}{r}
1{,}564\\
\times\ \ 5\\
\hline
\end{array}
$$

Estimate. Then find the product.

7. 4×979

8. 503×7

9. $5 \times 4{,}257$

10. $6{,}018 \times 9$

11. 758×6

12. 3×697

13. $2{,}141 \times 8$

14. $7 \times 7{,}956$

Problem Solving REAL WORLD

15. Mr. and Mrs. Dorsey and their three children are flying to Springfield. The cost of each ticket is $179. Estimate how much the tickets will cost. Then find the exact cost of the tickets.

_____

_____

_____

16. Ms. Tao flies roundtrip twice yearly between Jacksonville and Los Angeles on business. The distance between the two cities is 2,150 miles. Estimate the distance she flies for both trips. Then find the exact distance.

_____

_____

_____

Lesson Check

1. Mr. Nielson works 154 hours each month. He works 8 months each year. How many hours does Mr. Nielson work each year?

   (A) 832 hours

   (B) 1,232 hours

   (C) 1,502 hours

   (D) 1,600 hours

2. Sasha lives 1,493 miles from her grandmother. One year, Sasha's family made 4 round trips to visit her grandmother. How many miles did they travel in all?

   (A) 5,972 miles

   (B) 8,944 miles

   (C) 11,944 miles

   (D) 15,944 miles

Spiral Review

3. Yuna missed 5 points out of 100 points on her math test. What decimal number represents the part of her math test that she answered correctly? (Grade 4)

   (A) 0.05

   (B) 0.50

   (C) 0.75

   (D) 0.95

4. Which symbol makes the statement true? (Grade 4)

   602,163 ◯ 620,163

   (A) >

   (B) <

   (C) =

   (D) ÷

5. The number below represents the number of fans that attended Chicago Cubs baseball games in 2008. What is this number written in standard form? (Lesson 1.2)

   $(3 \times 1,000,000) + (3 \times 100,000) + (2 \times 100)$

   (A) 33,300,200

   (B) 30,300,200

   (C) 3,300,200

   (D)   330,200

6. A fair was attended by 755,082 people altogether. What is this number rounded to the nearest ten thousand? (Grade 4)

   (A) 800,000

   (B) 760,000

   (C) 750,000

   (D) 700,000

Name _____

Multiply by 2-Digit Numbers

Estimate. Then find the product.

1. Estimate: ____**4,000**____

```
     82
  ×  49
    738
 + 3280
  4,018
```

2. Estimate: _____

```
     92
  ×  68
```

3. Estimate: _____

```
    396
  ×  37
```

4. 23 × 67

5. 86 × 33

6. 78 × 71

_____

_____

_____

7. 309 × 29

8. 612 × 87

9. 476 × 72

_____

_____

_____

Problem Solving REAL WORLD

10. A company shipped 48 boxes of canned dog food. Each box contains 24 cans. How many cans of dog food did the company ship in all?

11. There were 135 cars in a rally. Each driver paid a $25 fee to participate in the rally. How much money did the drivers pay in all?

_____

_____

Lesson Check

1. A chessboard has 64 squares. At a chess tournament 84 chessboards were used. How many squares are there on 84 chessboards?

   (A) 4,816

   (B) 5,036

   (C) 5,166

   (D) 5,376

2. Last month, a manufacturing company shipped 452 boxes of ball bearings. Each box contains 48 ball bearings. How many ball bearings did the company ship last month?

   (A) 21,296

   (B) 21,686

   (C) 21,696

   (D) 21,706

Spiral Review

3. What is the standard form of the number three million, sixty thousand, five hundred twenty? (Lesson 1.2)

   (A) 3,060,520

   (B) 3,065,020

   (C) 3,600,520

   (D) 3,652,000

4. What number completes the following equation? (Lesson 1.3)

   $8 \times (40 + 7) = (8 \times \boxed{}) + (8 \times 7)$

   (A) 40

   (B) 47

   (C) 320

   (D) 376

5. The population of Clarksville is about 6,000 people. What is Clarksville's population written as a whole number multiplied by a power of ten? (Lesson 1.4)

   (A) 6×10^1

   (B) 6×10^2

   (C) 6×10^3

   (D) 6×10^4

6. A sporting goods store ordered 144 cans of tennis balls. Each can contains 3 balls. How many tennis balls did the store order?

   (Lesson 1.6)

   (A) 342

   (B) 412

   (C) 422

   (D) 432

Name _____

Relate Multiplication to Division

Use multiplication and the Distributive Property to find the quotient.

1. $70 \div 5 =$ _____**14**_____

$(5 \times 10) + (5 \times 4) = 70$

$5 \times 14 = 70$

2. $96 \div 6 =$ _____

3. $85 \div 5 =$ _____

4. $84 \div 6 =$ _____

5. $168 \div 7 =$ _____

6. $104 \div 4 =$ _____

7. $171 \div 9 =$ _____

8. $102 \div 6 =$ _____

9. $210 \div 5 =$ _____

Problem Solving REAL WORLD

10. Ken is making gift bags for a party. He has 64 colored pens and wants to put the same number in each bag. How many bags will Ken make if he puts 4 pens in each bag?

_____

11. Maritza is buying wheels for her skateboard shop. She ordered a total of 92 wheels. If wheels come in packages of 4, how many packages will she receive?

_____

© Houghton Mifflin Harcourt Publishing Company

Lesson Check

1. Which of the following expressions can be used to find $36 \div 3$?

    Ⓐ $(3 \times 10) + (3 \times 2)$

    Ⓑ $(6 \times 10) + (6 \times 2)$

    Ⓒ $(3 \times 12) + (3 \times 2)$

    Ⓓ $(2 \times 10) + (3 \times 12)$

2. Which of the following expressions can be used to find $126 \div 7$?

    Ⓐ $(7 \times 20) + (7 \times 6)$

    Ⓑ $(7 \times 10) + (7 \times 8)$

    Ⓒ $(6 \times 20) + (6 \times 1)$

    Ⓓ $(2 \times 50) + (2 \times 13)$

Spiral Review

3. Allison separates her 23 stickers into 4 equal piles. How many stickers does she have left over? **(Grade 4)**

    Ⓐ 27

    Ⓑ 19

    Ⓒ 5

    Ⓓ 3

4. A website had 2,135,789 hits. What is the value of the digit 3? **(Lesson 1.2)**

    Ⓐ   30

    Ⓑ   3,000

    Ⓒ   30,000

    Ⓓ 300,000

5. The area of Arizona is 114,006 square miles. What is the expanded form of this number?
    (Lesson 1.2)

    Ⓐ $(1 \times 100,000) + (1 \times 1,400) + (6 \times 1)$

    Ⓑ $(1 \times 100,000) + (1 \times 11,000) + (1 \times 4,000) + (6 \times 1)$

    Ⓒ $(1 \times 100,000) + (1 \times 10,000) + (4 \times 1,000) + (6 \times 1)$

    Ⓓ $(1 \times 11,000) + (1 \times 4,000) + (6 \times 1)$

6. Which of the following shows the value of the fourth power of ten? **(Lesson 1.4)**

    Ⓐ   1,000

    Ⓑ   10,000

    Ⓒ   100,000

    Ⓓ 1,000,000

Name _____

Problem Solving • Multiplication and Division

Solve the problems below. Show your work.

1. Dani is making punch for a family picnic. She adds 16 fluid ounces of orange juice, 16 fluid ounces of lemon juice, and 8 fluid ounces of lime juice to 64 fluid ounces of water. How many 8-ounce glasses of punch can she fill?

$$104 \div 8 = (40 + 64) \div 8$$
$$= (40 \div 8) + (64 \div 8)$$
$$= 5 + 8, \text{ or } 13$$

16 + 16 + 8 + 64 = 104 fluid ounces

13 glasses

2. Ryan has nine 14-ounce bags of popcorn to repackage and sell at the school fair. A small bag holds 3 ounces. How many small bags can he make?

_____

3. Bianca is making scarves to sell. She has 33 pieces of blue fabric, 37 pieces of green fabric, and 41 pieces of red fabric. Suppose Bianca uses 3 pieces of fabric to make 1 scarf. How many scarves can she make?

_____

4. Jasmine has 8 packs of candle wax to make scented candles. Each pack contains 14 ounces of wax. Jasmine uses 7 ounces of wax to make one candle. How many candles can she make?

_____

5. Maurice puts 130 trading cards in protector sheets. He fills 7 sheets and puts the remaining 4 cards in an eighth sheet. Each of the filled sheets has the same number of cards. How many cards are in each filled sheet?

_____

Lesson Check

1. Joyce is helping her aunt create craft kits. Her aunt has 138 pipe cleaners, and each kit will include 6 pipe cleaners. How many kits can they make?

   (A) 13
   (B) 18
   (C) 22
   (D) 23

2. Stefan plants seeds for 30 carrot plants and 45 beet plants in 5 rows, with the same number of seeds in each row. How many seeds are planted in each row?

   (A) 10
   (B) 14
   (C) 15
   (D) 80

Spiral Review

3. Georgia wants to evenly divide 84 trading cards between 6 friends. How many cards will each friend get? (Lesson 1.8)

   (A) 12
   (B) 13
   (C) 14
   (D) 16

4. Maria has 144 marbles. Emanuel has 4 times the number of marbles Maria has. How many marbles does Emanuel have? (Lesson 1.6)

   (A) 36
   (B) 140
   (C) 566
   (D) 576

5. The Conservation Society bought and planted 45 cherry trees. Each tree cost $367. What was the total cost of planting the trees? (Lesson 1.7)

   (A) $3,303
   (B) $16,485
   (C) $16,515
   (D) $20,185

6. A sports arena covers 710,430 square feet of ground. A newspaper reported that the arena covers about 700,000 square feet of ground. To what place value was the number rounded? (Grade 4)

   (A) hundreds
   (B) thousands
   (C) ten thousands
   (D) hundred thousands

Name _____

Numerical Expressions

Write an expression to match the words.

1. Ethan collected 16 seashells. He lost 4 of them while walking home.

$$16 - 4$$
_____

2. Yasmine bought 4 bracelets. Each bracelet cost $3.

_____

3. Amani did 10 jumping jacks. Then she did 7 more.

_____

4. Darryl has a board that is 8 feet long. He cuts it into pieces that are each 2 feet long.

_____

Write words to match the expression.

5. $3 + (4 \times 12)$

_____
_____

6. $36 \div 4$

_____
_____

7. $24 - (6 + 3)$

_____
_____

Draw a line to match the expression with the words.

8. Ray picked 30 apples and put them equally into 3 baskets. Then he ate two of the apples in a basket.

9. Quinn had $30. She bought a notebook for $3 and a pack of pens for $2.

10. Colleen runs 3 miles twice a day for 30 days.

$(3 \times 2) \times 30$

$(30 \div 3) - 2$

$30 - (3 + 2)$

Problem Solving  REAL WORLD

11. Kylie has 14 polished stones. Her friend gives her 6 more stones. Write an expression to match the words.

_____

12. Rashad had 25 stamps. He shared them equally among himself and 4 friends. Then Rashad found 2 more stamps in his pocket. Write an expression to match the words.

_____

Lesson Check

1. Jenna bought 3 packs of bottled water, with 8 bottles in each pack. Then she gave 6 bottles away. Which expression matches the words?

   Ⓐ (3 + 8) + 6

   Ⓑ (3 × 8) × 6

   Ⓒ (3 × 8) + 6

   Ⓓ (3 × 8) − 6

2. Stephen had 24 miniature cars. He gave 4 cars to his brother. Then he passed the rest of the cars out equally among 4 of his friends. Which operation would you use to represent the first part of this situation?

   Ⓐ addition

   Ⓑ subtraction

   Ⓒ division

   Ⓓ multiplication

Spiral Review

3. To find 36 + 29 + 14, Joshua rewrote the expression as 36 + 14 + 29. What property did Joshua use to rewrite the expression? (Lesson 1.3)

   Ⓐ Commutative Property of Multiplication

   Ⓑ Commutative Property of Addition

   Ⓒ Associative Property of Addition

   Ⓓ Associative Property of Multiplication

4. There are 6 baskets on the table. Each basket has 144 crayons in it. How many crayons are there in all? (Lesson 1.6)

   Ⓐ 644

   Ⓑ 664

   Ⓒ 844

   Ⓓ 864

5. Mr. Anderson wrote $(7 \times 9) \times 10^3$ on the board. What is the value of that expression? (Lesson 1.5)

   Ⓐ 630

   Ⓑ 6,300

   Ⓒ 63,000

   Ⓓ 630,000

6. Barbara mixes 54 ounces of granola and 36 ounces of raisins. She divides the mixture into 6-ounce servings. How many servings does she make? (Lesson 1.9)

   Ⓐ 3

   Ⓑ 12

   Ⓒ 15

   Ⓓ 96

Name _____

Evaluate Numerical Expressions

Evaluate the numerical expression.

1. $24 \times 5 - 41$

$120 - 41$

_____ **79** _____

2. $(32 - 20) \div 4$

_____

3. $16 \div (2 + 6)$

_____

4. $15 \times (8 - 3)$

_____

5. $4 \times 8 - 7$

_____

6. $27 + 5 \times 6$

_____

7. $3 \div 3 \times 4 + 6$

_____

8. $14 + 4 \times 4 - 9$

_____

Rewrite the expression with parentheses to equal the given value.

9. $3 \times 4 - 1 + 2$

value: 11

_____

10. $2 \times 6 \div 2 + 1$

value: 4

_____

11. $5 + 3 \times 2 - 6$

value: 10

_____

Problem Solving REAL WORLD

12. Sandy has several pitchers to hold lemonade for the school bake sale. Two pitchers can hold 64 ounces each, and four pitchers can hold 48 ounces each. How many total ounces can Sandy's pitchers hold?

_____

13. At the bake sale, Jonah sold 4 cakes for $8 each and 36 muffins for $2 each. What was the total amount, in dollars, that Jonah received from these sales?

_____

Lesson Check

1. What is the value of the expression
   $4 \times (4 - 2) + 6$?

   Ⓐ 6

   Ⓑ 14

   Ⓒ 24

   Ⓓ 40

2. Lannie ordered 12 copies of the same book
   for his book club members. The books cost
   $19 each, and the order has a $15 shipping
   charge. What is the total cost of Lannie's
   order?

   Ⓐ $243

   Ⓑ $213

   Ⓒ $199

   Ⓓ $161

Spiral Review

3. A small company packs 12 jars of jelly into
   each of 110 boxes to bring to the farmers'
   market. How many jars of jelly does the
   company pack in all? (Lesson 1.7)

   Ⓐ 1,220

   Ⓑ 1,320

   Ⓒ 1,350

   Ⓓ 2,300

4. June has 42 sports books, 85 mystery books,
   and 69 nature books. She arranges her
   books equally on 7 shelves. How many books
   are on each shelf? (Lesson 1.9)

   Ⓐ 12

   Ⓑ 18

   Ⓒ 28

   Ⓓ 196

5. Last year, a widget factory produced one
   million, twelve thousand, sixty widgets. What
   is this number written in standard form?
   (Lesson 1.2)

   Ⓐ 1,012,060

   Ⓑ 1,012,600

   Ⓒ 1,120,060

   Ⓓ 112,000,060

6. A company has 3 divisions. Last year, each
   division earned a profit of 5×10^5. What
   was the total profit the company earned last
   year? (Lesson 1.4)

   Ⓐ $50,000

   Ⓑ $150,000

   Ⓒ $500,000

   Ⓓ $1,500,000

Name _____

Grouping Symbols

Evaluate the numerical expression.

1. $5 \times [(11 - 3) - (13 - 9)]$

$5 \times [8 - (13 - 9)]$
$5 \times [8 - 4]$
5×4

_____ **20**

2. $30 - [(9 \times 2) - (3 \times 4)]$

3. $36 \div [(14 - 5) - (10 - 7)]$

4. $7 \times [(9 + 8) - (12 - 7)]$

5. $[(25 - 11) + (15 - 9)] \div 5$

6. $[(8 \times 9) - (6 \times 7)] - 15$

7. $8 \times \{[(7 + 4) \times 2] - [(11 - 7) \times 4]\}$

8. $\{[(8 - 3) \times 2] + [(5 \times 6) - 5]\} \div 5$

Problem Solving ~REAL WORLD~

Use the information at the right for 9 and 10.

9. Write an expression to represent the total number of muffins and cupcakes Joan sells in 5 days.

Joan has a cafe. Each day, she bakes 24 muffins. She gives away 3 and sells the rest. Each day, she also bakes 36 cupcakes. She gives away 4 and sells the rest.

_____

10. Evaluate the expression to find the total number of muffins and cupcakes Joan sells in 5 days.

_____

Lesson Check

1. What is the value of the expression?

   $9 \times [(21 - 4) - (2 + 7)]$

   (A) 72

   (B) 108

   (C) 190

   (D) 198

2. Which expression has a value of 24?

   (A) $[(17 - 9) \times (3 + 2)] \div 2$

   (B) $[(17 + 9) - (3 + 2)] - 2$

   (C) $[(17 - 9) \times (3 \times 2)] \div 2$

   (D) $[(17 - 9) + (3 \times 2)] \times 2$

Spiral Review

3. What is $\frac{1}{10}$ of 200? **(Lesson 1.1)**

   (A)   2

   (B)   20

   (C) 2,000

   (D) 20,000

4. The Park family is staying at a hotel near an amusement park for 3 nights. The hotel costs $129 per night. How much will their 3-night stay in the hotel cost? **(Lesson 1.6)**

   (A) $67

   (B) $369

   (C) $378

   (D) $387

5. Vidal bought 2 pizzas and cut each into 8 slices. He and his friends ate 10 slices. Which expression matches the words?

   (Lesson 1.10)

   (A) $(2 + 8) - 10$

   (B) $(2 \times 8) - 10$

   (C) $(2 \times 8) + 10$

   (D) $(2 \times 10) - 8$

6. What is the value of the underlined digit in 783,5<u>4</u>9,201? **(Lesson 1.2)**

   (A)   4

   (B)   40

   (C)   40,000

   (D) 400,000

Chapter 1 Extra Practice

Lessons 1.1 - 1.2

Complete the sentence.

1. 300 is 10 times as much as _____ .

2. 400 is $\frac{1}{10}$ of _____ .

Write the value of the underlined digit.

3. 4<u>5</u>,130

4. 8,<u>1</u>23,476

5. 153,<u>4</u>71

6. 6,5<u>8</u>3,450

_____   _____   _____   _____

Lesson 1.3

Complete the equation, and tell which property you used.

1. $(18 \times 2) \times 5 = 18 \times (2 \times$ _____ $)$

2. $64 + 58 =$ _____ $+ 64$

_____   _____

_____   _____

Lessons 1.4 - 1.5

Find the value.

1. 10^2

2. 10^5

3. 6×10^3

4. 8×10^7

_____   _____   _____   _____

5. $(6 \times 7) \times 10^3$

6. $(5 \times 4) \times 10^2$

7. $(3 \times 9) \times 10^6$

8. $(5 \times 8) \times 10^0$

_____   _____   _____   _____

Lessons 1.6 - 1.7

Estimate. Then find the product.

1. Estimate _____

429
× 5

2. Estimate _____

1,785
× 8

3 Estimate _____

81
× 22

4. Estimate _____

558
× 44

5. 9×802

6. $3,699 \times 7$

7. 34×93

8. 678×87

_____   _____   _____   _____

Lessons 1.8 - 1.9

Solve.

1. Morton and three of his friends earned a total of $168 walking dogs. They want to share the money equally. How much will each person get?

2. To make fruit salad, Sara uses 28 ounces of pineapple, 21 ounces of apples, 19 ounces of bananas, and 16 ounces of mango. How many 6-ounce servings of fruit salad can Sarah make?

Lesson 1.10

Write an expression to match the words.

1. Marilyn has 8 pears. She eats 2 pears.

2. Lee spends $9 on 3 comic books. Each comic book costs the same amount.

3. Al bought 24 stickers. He gave away 11 stickers. Then he bought 8 more stickers.

4. Nicky has 4 boxes of markers. Each box contains 8 markers.

Lessons 1.11 - 1.12

Evaluate the numerical expression.

1. $3 + 4 \times 6$

2. $8 - 2 \times 3$

3. $5 \div 5 \times 7 + 1$

4. $8 + 56 - 8 \times 4$

5. $12 - (3 + 4)$

6. $[18 \div (2 \times 3)] \times 4$

7. $24 - \{[16 - (8 - 1)] \times 2\}$

School-Home Letter

© Houghton Mifflin Harcourt Publishing Company

Vocabulary

compatible numbers Numbers that are easy to compute with mentally

dividend The number that is to be divided in a division problem

divisor The number that divides the dividend

quotient The number, not including the remainder, that results from dividing

remainder The amount left over when a number cannot be divided equally

Dear Family,

During the next few weeks, our math class will be learning about dividing three- and four-digit whole numbers. We will also learn how to interpret remainders.

You can expect to see homework that provides practice with division of three- and four-digit dividends by one- and two-digit divisors.

Here is a sample of how your child will be taught to divide a three-digit number by a one-digit divisor.

🔑 MODEL Divide Three-Digit Numbers

This is how we will divide three-digit numbers.

Solve. 268 ÷ 5

STEP 1

Estimate to place the first digit in the quotient.

250 ÷ 5 = 50

So, place the first digit in the tens place.

$$\begin{array}{r} 5 \\ 5\overline{)268} \end{array}$$

STEP 2

Divide the tens.

$$\begin{array}{r} 5 \\ 5\overline{)268} \\ -25 \\ \hline 18 \end{array}$$

STEP 3

Divide the ones.

$$\begin{array}{r} 53\text{ r}3 \\ 5\overline{)268} \\ -25 \\ \hline 18 \\ -15 \\ \hline 3 \end{array}$$

Tips

Identifying Patterns in Division

When estimating to place the first digit, it is important to recognize patterns with multiples of 10, 100, and 1,000. Complete the division with basic facts, then attach the same number of zeros to the dividend and the quotient.

36 ÷ 4 = 9
so, 36,000 ÷ 4 = 9,000.

Activity

Plan a vacation for the summer. Research the distance to the destination from your home. You can spend no more than one week traveling to the destination, and you must travel the same number of miles each day. Decide how many days you will spend traveling. Then find how many miles you need to travel each day.

Carta para la casa

Querida familia,

Durante las próximas semanas en la clase de matemáticas aprenderemos a dividir números enteros de tres y cuatro dígitos. También aprenderemos a interpretar los residuos.

Llevaré a la casa tareas con actividades para practicar la división de dividendos de tres y cuatro dígitos entre divisores de uno y dos dígitos.

Este es un ejemplo de la manera como aprenderemos a dividir un número de tres dígitos entre un número de un dígito.

Vocabulario

números compatibles Números con los que se pueden hacer cálculos mentales fácilmente

dividendo El número que se va a dividir en un problema de división

divisor El número entre el que se divide el dividendo

cociente El número, sin incluir al residuo, que resulta de la división

residuo La cantidad que sobra cuando un número no se puede dividir en partes iguales

🔑 MODELO Dividir números de tres dígitos

Este es un ejemplo de cómo dividir números de tres dígitos.

Resuelve $268 \div 5$

PASO 1

Estima para colocar el primer dígito del cociente.

$250 \div 5 = 50$

Por lo tanto, coloca el primer dígito en el lugar de las decenas.

$$5 \overline{)268}$$

PASO 2

Divide las decenas.

$$\begin{array}{r} 5 \\ 5\overline{)268} \\ -25 \\ \hline 18 \end{array}$$

PASO 3

Divide las unidades.

$$\begin{array}{r} 53\ r3 \\ 5\overline{)268} \\ -25 \\ \hline 18 \\ -15 \\ \hline 3 \end{array}$$

Pistas

Identificar patrones en la división

Al estimar para colocar el primer dígito, es importante identificar patrones con múltiplos de 10, 100 y 1000. Completa la división con operaciones básicas, luego agrega el mismo número de ceros al dividendo y al cociente.

$$36 \div 4 = 9$$

Por lo tanto,

$$36,000 \div 4 = 9,000$$

Actividad

Planeen unas vacaciones de verano. Investiguen cuál es la distancia desde su casa hasta el destino. Sólo tienen una semana para hacer el viaje y deben recorrer el mismo número de millas cada día. Decidan cuántas millas deben recorrer cada día.

Name _____

Place the First Digit

Divide.

1. 4)‾388‾

$$
\begin{array}{r}
97 \\
4{\overline{\smash{\big)}\,388}} \\
-36 \\
\hline
28 \\
-28 \\
\hline
0
\end{array}
$$

97

2. 4)‾457‾

3. 8)‾712‾

4. 9)‾204‾

5. 2,117 ÷ 3

6. 520 ÷ 8

7. 1,812 ÷ 4

8. 3,476 ÷ 6

Problem Solving REAL WORLD

9. The school theater department made $2,142 on ticket sales for the three nights of their play. The department sold the same number of tickets each night and each ticket cost $7. How many tickets did the theater department sell each night?

10. Andreus made $625 mowing yards. He worked for 5 consecutive days and earned the same amount of money each day. How much money did Andreus earn per day?

Lesson Check

1. Kenny is packing cans into bags at the food bank. He can pack 8 cans into each bag. How many bags will Kenny need for 1,056 cans?

   (A) 133

   (B) 132

   (C) 131

   (D) 130

2. Liz polishes rings for a jeweler. She can polish 9 rings per hour. How many hours will it take her to polish 315 rings?

   (A) 45 hours

   (B) 35 hours

   (C) 25 hours

   (D) 15 hours

Spiral Review

3. Fiona uses 256 fluid ounces of juice to make 1 bowl of punch. How many fluid ounces of juice will she use to make 3 bowls of punch? (Lesson 1.7)

   (A) 56 fluid ounces

   (B) 128 fluid ounces

   (C) 512 fluid ounces

   (D) 768 fluid ounces

4. Len wants to write the number 100,000 using a base of 10 and an exponent. What number should he use as the exponent? (Lesson 1.4)

   (A) 4

   (B) 5

   (C) 10

   (D) 100,000

5. Family passes to an amusement park cost $54 each. Which expression can be used to find the cost in dollars of 8 family passes? (Lesson 1.3)

   (A) $(8 + 50) + (8 + 4)$

   (B) $(8 + 50) \times (8 + 4)$

   (C) $(8 \times 50) + (8 \times 4)$

   (D) $(8 \times 50) \times (8 \times 4)$

6. Gary is catering a picnic. There will be 118 guests at the picnic, and he wants each guest to have a 12-ounce serving of salad. How much salad should he make? (Lesson 1.7)

   (A) 216 ounces

   (B) 1,180 ounces

   (C) 1,416 ounces

   (D) 1,516 ounces

Name _____

Divide by 1-Digit Divisors

Divide.

1. $4\overline{)724}$

2. $5\overline{)312}$

3. $278 \div 2$

4. $336 \div 7$

$$
\begin{array}{r}
181 \\
4\overline{)724} \\
-4 \\
\hline
32 \\
-32 \\
\hline
04 \\
-\ 4 \\
\hline
0
\end{array}
$$

181 _____   _____   _____

Find the value of *n* in each equation. Write what *n* represents
in the related division problem.

5. $n = 3 \times 45$

6. $643 = 4 \times 160 + n$

7. $n = 6 \times 35 + 4$

_____   _____   _____

Problem Solving REAL WORLD

8. Randy has 128 ounces of dog food. He feeds
his dog 8 ounces of food each day. How many
days will the dog food last?

9. Angelina bought a 64-ounce can of lemonade
mix. She uses 4 ounces of mix for each pitcher
of lemonade. How many pitchers of lemonade
can Angelina make from the can of mix?

_____   _____

Lesson Check

1. A color printer will print 8 pages per minute. How many minutes will it take to print a report that has 136 pages?

Ⓐ 18 minutes

Ⓑ 17 minutes

Ⓒ 16 minutes

Ⓓ 15 minutes

2. A postcard collector has 1,230 postcards. If she displays them on pages that hold 6 cards each, how many pages does she need?

Ⓐ 230

Ⓑ 215

Ⓒ 205

Ⓓ 125

Spiral Review

3. Francis is buying a stereo system for $196. She wants to pay for it in four equal monthly installments. What is the amount she will pay each month? **(Lesson 2.1)**

Ⓐ $39

Ⓑ $49

Ⓒ $192

Ⓓ $784

4. A bakery bakes 184 loaves of bread in 4 hours. How many loaves does the bakery bake in 1 hour? **(Lesson 2.2)**

Ⓐ 736

Ⓑ 180

Ⓒ 92

Ⓓ 46

5. Marvin collects trading cards. He stores them in boxes that hold 235 cards each. If Marvin has 4 boxes full of cards, how many cards does he have in his collection?

(Lesson 1.6)

Ⓐ 940

Ⓑ 920

Ⓒ 800

Ⓓ 705

6. Which number has the digit 7 in the ten-thousands place? **(Lesson 1.2)**

Ⓐ 810,745

Ⓑ 807,150

Ⓒ 708,415

Ⓓ 870,541

Name _____

Division with 2-Digit Divisors

Use the quick picture to divide.

1. $132 \div 12 =$ __**11**__

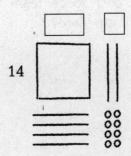

2. $168 \div 14 =$ _____

Divide. Use base-ten blocks.

3. $195 \div 13 =$ _____

4. $143 \div 11 =$ _____

5. $165 \div 15 =$ _____

Divide. Draw a quick picture.

6. $192 \div 16 =$ _____

7. $169 \div 13 =$ _____

Problem Solving

8. There are 182 seats in a theater. The seats are evenly divided into 13 rows. How many seats are in each row?

9. There are 156 students at summer camp. The camp has 13 cabins. An equal number of students sleep in each cabin. How many students sleep in each cabin?

_____

_____

Lesson Check

1. There are 198 students in the soccer league. There are 11 players on each soccer team. How many soccer teams are there?

   Ⓐ 17
   Ⓑ 18
   Ⓒ 19
   Ⓓ 21

2. Jason earned $187 for 17 hours of work. How much did Jason earn per hour?

   Ⓐ $11
   Ⓑ $12
   Ⓒ $13
   Ⓓ $14

Spiral Review

3. Which number represents six million, seven hundred thousand, twenty? (Lesson 1.2)

   Ⓐ 6,000,720
   Ⓑ 6,007,020
   Ⓒ 6,700,020
   Ⓓ 6,720,000

4. Which expression represents the sentence "Add the product of 3 and 6 to 4?" (Lesson 1.10)

   Ⓐ (4 + 3) × 6
   Ⓑ 4 + (3 × 6)
   Ⓒ 4 + (3 + 6)
   Ⓓ 4 × 3 + 6

5. To transport 228 people to an island, the island ferry makes 6 different trips. On each trip, the ferry carries the same number of people. How many people does the ferry transport on each trip? (Lesson 2.2)

   Ⓐ 36
   Ⓑ 37
   Ⓒ 38
   Ⓓ 39

6. Isabella sells 36 tickets to the school talent show. Each ticket costs $14. How much money does Isabella collect for the tickets she sells? (Lesson 1.7)

   Ⓐ $180
   Ⓑ $384
   Ⓒ $404
   Ⓓ $504

Name _____

Partial Quotients

Divide. Use partial quotients.

1. $18\overline{)236}$

$$
\begin{array}{r}
18\overline{)236} \\
-180 \leftarrow 10 \times 18 \quad 10 \\
\hline
56 \\
-36 \leftarrow 2 \times 18 \quad 2 \\
\hline
20 \\
-18 \leftarrow 1 \times 18 \quad +1 \\
\hline
2 \qquad\qquad 13
\end{array}
$$

236 ÷ 18 is 13 r2.

2. $36\overline{)540}$

3. $27\overline{)624}$

4. $478 \div 16$

5. $418 \div 22$

6. $625 \div 25$

7. $514 \div 28$

8. $322 \div 14$

9. $715 \div 25$

Problem Solving REAL WORLD

10. A factory processes 1,560 ounces of olive oil per hour. The oil is packaged into 24-ounce bottles. How many bottles does the factory fill in one hour?

_____

11. A pond at a hotel holds 4,290 gallons of water. The groundskeeper drains the pond at a rate of 78 gallons of water per hour. How long will it take to drain the pond?

_____

Lesson Check

1. Yvette has 336 eggs to put into cartons. She puts one dozen eggs into each carton. How many cartons does she fill?

   Ⓐ 20
   Ⓑ 21
   Ⓒ 27
   Ⓓ 28

2. Ned mows a 450 square-foot garden in 15 minutes. How many square feet of the garden does he mow in one minute?

   Ⓐ 3 square feet
   Ⓑ 30 square feet
   Ⓒ 435 square feet
   Ⓓ 465 square feet

Spiral Review

3. Raul has 56 bouncy balls. He puts three times as many balls into red gift bags as he puts into green gift bags. If he puts the same number of balls into each bag, how many balls does he put into green bags? (Lesson 1.9)

   Ⓐ 42
   Ⓑ 19
   Ⓒ 14
   Ⓓ 12

4. Marcia uses 5 ounces of chicken stock to make one batch of soup. She has a total of 400 ounces of chicken stock. How many batches of soup can Marcia make? (Lesson 2.2)

   Ⓐ 50
   Ⓑ 80
   Ⓒ 200
   Ⓓ 2,000

5. Michelle buys 13 bags of gravel for her fish aquarium. If each bag weighs 12 pounds, how many pounds of gravel did she buy? (Lesson 1.7)

   Ⓐ 156 pounds
   Ⓑ 143 pounds
   Ⓒ 130 pounds
   Ⓓ 26 pounds

6. Which of the following represents 4,305,012 in expanded notation? (Lesson 1.2)

   Ⓐ 400,000 + 30,000 + 5,000 + 12
   Ⓑ 40,000 + 3,000 + 500 + 10 + 2
   Ⓒ 4,000,000 + 300,000 + 5,000 + 100 + 2
   Ⓓ 4,000,000 + 300,000 + 5,000 + 10 + 2

Name _____

Estimate with 2-Digit Divisors

Use compatible numbers to find two estimates.

1. $18\overline{)1{,}322}$ 2. $17\overline{)1{,}569}$ 3. $27\overline{)735}$ 4. $12\overline{)478}$

$$1{,}200 \div 20$$
$$= 60$$
$$1{,}400 \div 20$$
$$= 70$$

5. $336 \div 12$ 6. $1{,}418 \div 22$ 7. $16\overline{)2{,}028}$ 8. $2{,}242 \div 33$

Use compatible numbers to estimate the quotient.

9. $82\overline{)5{,}514}$ 10. $61\overline{)5{,}320}$ 11. $28\overline{)776}$ 12. $23\overline{)1{,}624}$

Problem Solving REAL WORLD

13. A cubic yard of topsoil weighs 4,128 pounds. About how many 50-pound bags of topsoil can you fill with one cubic yard of topsoil?

14. An electronics store places an order for 2,665 USB flash drives. One shipping box holds 36 flash drives. About how many boxes will it take to hold all the flash drives?

_____   _____

Lesson Check

1. Marcy has 567 earmuffs in stock. If she can put 18 earmuffs on each shelf, about how many shelves does she need for all the earmuffs?

   (A) about 20

   (B) about 30

   (C) about 570

   (D) about 590

2. Howard pays $327 for one dozen collector's edition baseball cards. About how much does he pay for each baseball card?

   (A) about $20

   (B) about $30

   (C) about $45

   (D) about $50

Spiral Review

3. Andrew can frame 9 pictures each day. He has an order for 108 pictures. How many days will it take him to complete the order? (Lesson 2.2)

   (A) 10 days

   (B) 11 days

   (C) 12 days

   (D) 13 days

4. Madeleine can type 3 pages in one hour. How many hours will it take her to type a 123-page report? (Lesson 2.2)

   (A) 65 hours

   (B) 45 hours

   (C) 41 hours

   (D) 22 hours

5. Suppose you round 43,257,529 to 43,300,000. To what place value did you round the number? (Lesson 1.2)

   (A) hundred thousands

   (B) ten thousands

   (C) thousands

   (D) tens

6. Grace's catering received an order for 118 apple pies. Grace uses 8 apples to make one apple pie. How many apples does she need to make all 118 pies? (Lesson 1.6)

   (A) 110

   (B) 126

   (C) 884

   (D) 944

Name _____

Divide by 2-Digit Divisors

Divide. Check your answer.

1. $385 \div 12$

$$
\begin{array}{r}
32 \text{ r1} \\
12\overline{)385} \\
-36 \\
\hline
25 \\
-24 \\
\hline
1
\end{array}
$$

2. $837 \div 36$

3. $1,650 \div 55$

4. $5,634 \div 18$

5. $7,231 \div 24$

6. $5,309 \div 43$

7. $37\overline{)3,774}$

8. $54\overline{)1,099}$

9. $28\overline{)6,440}$

10. $52\overline{)5,256}$

11. $85\overline{)1,955}$

12. $46\overline{)5,624}$

Problem Solving  REAL WORLD

13. The factory workers make 756 machine parts in 36 hours. Suppose the workers make the same number of machine parts each hour. How many machine parts do they make each hour?

14. One bag holds 12 bolts. Several bags filled with bolts are packed into a box and shipped to the factory. The box contains a total of 2,760 bolts. How many bags of bolts are in the box?

_____

Lesson Check

1. A bakery packages 868 cupcakes into 31 boxes. The same number of cupcakes are put into each box. How many cupcakes are in each box?

   (A) 28

   (B) 37

   (C) 38

   (D) 47

2. Maggie orders 19 identical gift boxes. The Ship-Shape Packaging Company packs and ships the boxes for $1,292. How much does it cost to pack and ship each box?

   (A) $58

   (B) $64

   (C) $68

   (D) $78

Spiral Review

3. What is the standard form of the number four million, two hundred sixteen thousand, ninety? (Lesson 1.2)

   (A) 4,260,090

   (B) 4,216,900

   (C) 4,216,090

   (D) 4,216,019

4. Kelly and 23 friends go roller skating. They pay a total of $186. About how much does it cost for one person to skate? (Lesson 2.5)

   (A) about $9

   (B) about $15

   (C) about $25

   (D) about $200

5. In two days, Gretchen drinks seven 16-ounce bottles of water. She drinks the water in 4 equal servings. How many ounces of water does Gretchen drink in each serving? (Lesson 1.9)

   (A) 112 ounces

   (B) 28 ounces

   (C) 12 ounces

   (D) 4 ounces

6. What is the value of the underlined digit in 5,4̲36,788? (Lesson 1.2)

   (A) 4

   (B) 400

   (C) 40,000

   (D) 400,000

Name _____

Interpret the Remainder

Interpret the remainder to solve.

1. Warren spent 140 hours making 16 wooden toy trucks for a craft fair. If he spent the same amount of time making each truck, how many hours did he spend making each truck?

$$
\begin{array}{r}
8 \\
16\overline{)140} \\
-128 \\
\hline
12
\end{array}
$$

$8\frac{3}{4}$ hours

_____

2. Marcia has 412 bouquets of flowers for centerpieces. She uses 8 flowers for each centerpiece. How many centerpieces can she make?

_____

3. On the 5th grade class picnic, 50 students share 75 sandwiches equally. How many sandwiches does each student get?

_____

4. One plant container holds 14 tomato seedlings. If you have 1,113 seedlings, how many containers do you need to hold all the seedlings?

_____

Problem Solving REAL WORLD

5. Fiona bought 212 stickers to make a sticker book. If she places 18 stickers on each page, how many pages will her sticker book have?

_____

6. Jenny has 220 ounces of cleaning solution that she wants to divide equally among 12 large containers. How much cleaning solution should she put in each container?

_____

Lesson Check

1. Henry and 28 classmates go to the roller skating rink. Each van can hold 11 students. If all of the vans are full except one, how many students are in the van that is not full?

    (A) 1

    (B) 2

    (C) 6

    (D) 7

2. Candy buys 20 ounces of mixed nuts. She puts an equal number of ounces in each of 3 bags. How many ounces of mixed nuts will be in each bag?

    (A) 7 ounces

    (B) $6\frac{2}{3}$ ounces

    (C) 6 ounces

    (D) 2 ounces

Spiral Review

3. Jayson earns $196 each week bagging groceries at the store. He saves half his earnings each week. How much money does Jayson save per week? (Lesson 2.2)

    (A) $92

    (B) $98

    (C) $102

    (D) $106

4. Desiree swims laps for 25 minutes each day. How many minutes does she spend swimming laps in 14 days? (Lesson 1.7)

    (A) 29 minutes

    (B) 125 minutes

    (C) 330 minutes

    (D) 350 minutes

5. Steve is participating in a bike-a-thon for charity. He will bike 144 miles per day for 5 days. How many miles will Steve bike in all? (Lesson 1.6)

    (A) 360 miles

    (B) 620 miles

    (C) 720 miles

    (D) 820 miles

6. Kasi is building a patio. He has 136 bricks. He wants the patio to have 8 rows, each with the same number of bricks. How many bricks will Kasi put in each row? (Lesson 2.2)

    (A) 17

    (B) 18

    (C) 128

    (D) 1,088

Name _____

Adjust Quotients

Adjust the estimated digit in the quotient, if needed. Then divide.

1.
$$\begin{array}{r} 5 \\ 16\overline{)976} \\ -80 \\ \hline 17 \end{array}$$

$$\begin{array}{r} 61 \\ 16\overline{)976} \\ -96 \\ \hline 16 \\ -16 \\ \hline 0 \end{array}$$

2.
$$\begin{array}{r} 3 \\ 24\overline{)689} \end{array}$$

3.
$$\begin{array}{r} 3 \\ 65\overline{)2,210} \end{array}$$

4.
$$\begin{array}{r} 2 \\ 38\overline{)7,035} \end{array}$$

Divide.

5. $2,961 \div 47$

6. $2,072 \div 86$

7. $1,280 \div 25$

8. $31\overline{)1,496}$

9. $86\overline{)6,290}$

10. $95\overline{)4,000}$

11. $44\overline{)2,910}$

12. $82\overline{)4,018}$

Problem Solving  REAL WORLD

13. A copier prints 89 copies in one minute. How long does it take the copier to print 1,958 copies?

14. Erica is saving her money to buy a dining room set that costs $580. If she saves $29 each month, how many months will she need to save to have enough money to buy the set?

Lesson Check

1. Gail ordered 5,675 pounds of flour for the bakery. The flour comes in 25-pound bags. How many bags of flour will the bakery receive?

   (A) 203

   (B) 227

   (C) 230

   (D) 5,650

2. Simone is in a bike-a-thon for a fundraiser. She receives $15 in pledges for every mile she bikes. If she wants to raise $510, how many miles does she need to bike?

   (A) 32 miles

   (B) 34 miles

   (C) 52 miles

   (D) 495 miles

Spiral Review

3. Lina makes beaded bracelets. She uses 9 beads to make each bracelet. How many bracelets can she make with 156 beads? (Lesson 2.2)

   (A) 12

   (B) 14

   (C) 17

   (D) 18

4. A total of 1,056 students from different schools enter the county science fair. Each school enters exactly 32 students. How many schools participate in the science fair? (Lesson 2.6)

   (A) 33,792

   (B) 352

   (C) 34

   (D) 33

5. Which number is less than 6,789,405? (Lesson 1.2)

   (A) 6,789,398

   (B) 6,789,599

   (C) 6,809,405

   (D) 6,879,332

6. Christy buys 48 barrettes. She shares the barrettes equally between herself and her 3 sisters. Which expression represents the number of barrettes each girl gets? (Lesson 1.10)

   (A) $\frac{48}{3} + 1$

   (B) $48 \div (3 + 1)$

   (C) $\frac{48}{3}$

   (D) $48 - 3$

Name _____

Problem Solving • Division

Show your work. Solve each problem.

1. Duane has 12 times as many baseball cards as Tony. Between them, they have 208 baseball cards. How many baseball cards does each boy have?

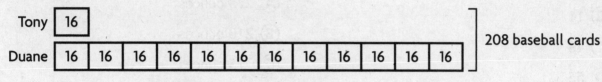

Tony | 16 |

Duane | 16 | 16 | 16 | 16 | 16 | 16 | 16 | 16 | 16 | 16 | 16 | 16 | ⎤ 208 baseball cards

$$208 \div 13 = 16$$

Tony: 16 cards; Duane: 192 cards

2. Hallie has 10 times as many pages to read for her homework assignment as Janet. Altogether, they have to read 264 pages. How many pages does each girl have to read?

3. Hank has 48 fish in his aquarium. He has 11 times as many tetras as guppies. How many of each type of fish does Hank have?

4. Kelly has 4 times as many songs on her music player as Lou. Tiffany has 6 times as many songs on her music player as Lou. Altogether, they have 682 songs on their music players. How many songs does Kelly have?

Lesson Check

1. Chelsea has 11 times as many art brushes as Monique. If they have 60 art brushes altogether, how many brushes does Chelsea have?

    (A) 5

    (B) 11

    (C) 49

    (D) 55

2. Jo has a gerbil and a German shepherd. The shepherd eats 14 times as much food as the gerbil. Altogether, they eat 225 ounces of dry food per week. How many ounces of food does the German shepherd eat per week?

    (A) 225 ounces

    (B) 210 ounces

    (C) 15 ounces

    (D) 14 ounces

Spiral Review

3. Jeanine is twice as old as her brother Marc. If the sum of their ages is 24, how old is Jeanine? (Lesson 1.9)

    (A) 16

    (B) 12

    (C) 8

    (D) 3

4. Larry is shipping nails that weigh a total of 53 pounds. He divides the nails equally among 4 shipping boxes. How many pounds of nails does he put in each box? (Lesson 2.7)

    (A) 4 pounds

    (B) 13 pounds

    (C) $13\frac{1}{4}$ pounds

    (D) 14 pounds

5. Annie plants 6 rows of small flower bulbs in a garden. She plants 132 bulbs in each row. How many bulbs does Annie plant in all?
    (Lesson 1.6)

    (A) 738

    (B) 792

    (C) 924

    (D) 984

6. Next year, four elementary schools will each send 126 students to Bedford Middle School. What is the total number of students the elementary schools will send to the middle school? (Lesson 1.6)

    (A) 46

    (B) 344

    (C) 444

    (D) 504

Name _____

Chapter 2 Extra Practice

Lessons 2.1 - 2.2

Divide.

1. $8\overline{)346}$

2. $6\overline{)1,914}$

3. $8\overline{)1,898}$

_____

_____

_____

4. $4\overline{)952}$

5. $3,629 \div 9$

6. $2,961 \div 7$

_____

_____

_____

7. $3\overline{)4,276}$

8. $6\overline{)3,251}$

9. $1,664 \div 5$

_____

_____

_____

Lessons 2.3 - 2.6

Estimate. Then divide.

1. $19\overline{)1,425}$

2. $2,384 \div 23$

3. $378 \div 56$

_____

_____

_____

Lesson 2.7

Interpret the remainder to solve.

1. Matthew is packing 195 glasses into cartons. Each carton can hold 18 glasses. How many cartons does Matthew need?

2. Julia wants to make 8 bows using 18 feet of ribbon. She wants to use an equal length of ribbon for each bow with no ribbon left over. How many feet of ribbon can she use for each bow?

_____

_____

Lesson 2.8

Divide.

1. $32\overline{)549}$

2. $1{,}296 \div 36$

3. $588 \div 84$

_____

_____

_____

4. $12\overline{)320}$

5. $53\overline{)6{,}681}$

6. $6{,}370 \div 29$

_____

_____

_____

Lesson 2.9

Solve.

1. Greenboro gets 12 times as many inches of snow as Redville gets. Altogether, the two towns get 65 inches of snow. How many inches of snow does Redville get?

2. In one month, Ansley runs 15 times as many miles as Zack runs. Altogether, they run 192 miles. How many miles does Ansley run in one month?

_____

_____

School-Home Letter

© Houghton Mifflin Harcourt Publishing Company

Dear Family,

Throughout the next few weeks, our math class will be studying decimals. We will be naming, comparing, ordering, and rounding decimals through thousandths. We will also be adding and subtracting decimals through hundredths.

You can expect to see homework that includes adding and subtracting decimals through hundredths.

Here is a sample of how your child will be taught to add decimals.

Vocabulary

decimal A number with one or more digits to the right of the decimal point

difference The result of subtracting two numbers

place value The value of each digit in a number based on the location of the digit

sum The result of adding two or more numbers

thousandth One of one thousand equal parts

🔑 MODEL Adding Decimals

Add 12.78 and 31.14.

STEP 1

Estimate the sum.

12.78 is about 13.

31.14 is about 31.

13 + 31 = 44

STEP 2

Write the problem with the decimal points aligned. Add the hundredths first. Then, add the tenths, ones, and tens. Regroup as needed.

```
   1
 12.78
+31.14
─────
 43.92
```

Tips

Adding and Subtracting Decimals

Always remember to align numbers on the decimal point when adding or subtracting decimals. That way, you are adding or subtracting the same place values.

Activity

Collect store advertisements from the newspaper. Have your child practice adding and subtracting decimals by writing and solving problems that involve money using the store advertisement.

Carta para la casa

Querida familia,

Durante las próximas semanas, en la clase de matemáticas estudiaremos los decimales. Nombraremos, compararemos, ordenaremos y redondearemos decimales hasta las milésimas. También sumaremos y restaremos decimales hasta las centésimas.

Llevaré a la casa tareas con actividades que incluyen sumar y restar decimales hasta los centésimos.

Este es un ejemplo de la manera como aprenderemos a sumar decimales.

🔑 MODELO Sumar decimales

Suma 12.78 y 31.14.

PASO 1

Estima la suma.

12.78 es aproximadamente 13.

31.14 es aproximadamente 31.

13 + 31 = 44

PASO 2

Escribe el problema con los puntos decimales alineados. Suma los centésimos primero. Después suma las décimas, las unidades y las decenas. Reagrupa si es necesario.

```
      1
   12.78
 +31.14
 ───────
   43.92
```

Pistas

No olvides alinear los números con el punto decimal cuando sumes o restes decimales. De esa manera estarás sumando y restando los mismos valores posicionales.

Actividad

Recorte algunos avisos publicitarios de varias tiendas que vea en el periódico. Pida a su hijo que use la información de los avisos para escribir y resolver problemas que incluyan cantidades de dinero.

Name _____

Thousandths

Write the decimal shown by the shaded parts of each model.

1.

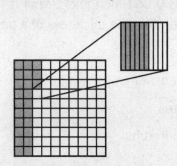

0.236

2.

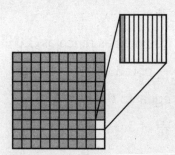

_____

Think: 2 tenths, 3 hundredths,
and 6 thousandths are shaded

Complete the sentence.

3. 0.4 is 10 times as much as _____ .

4. 0.003 is $\frac{1}{10}$ of _____ .

Use place-value patterns to complete the table.

Decimal	10 times as much as	$\frac{1}{10}$ of
5. 0.1		
6. 0.09		
7. 0.04		
8. 0.6		

Decimal	10 times as much as	$\frac{1}{10}$ of
9. 0.08		
10. 0.2		
11. 0.5		
12. 0.03		

Problem Solving REAL WORLD

13. The diameter of a dime is seven hundred five thousandths of an inch. Complete the table by recording the diameter of a dime.

14. What is the value of the 5 in the diameter of a half dollar?

_____

15. Which coins have a diameter with a 5 in the hundredths place?

_____

U.S. Coins	
Coin	**Diameter (in inches)**
Penny	0.750
Nickel	0.835
Dime	
Quarter	0.955
Half dollar	1.205

Lesson Check

1. What is the relationship between 3.0 and 0.3?

(A) 0.3 is 10 times as much as 3.0

(B) 3.0 is $\frac{1}{10}$ of 0.3

(C) 3.0 is equal to 0.3

(D) 0.3 is $\frac{1}{10}$ of 3.0

2. A penny is 0.061 inch thick. What is the value of the 6 in the thickness of a penny?

(A) 6 tens

(B) 6 thousandths

(C) 6 tenths

(D) 6 hundredths

Spiral Review

3. What is the number seven hundred thirty-one million, nine hundred thirty-four thousand, thirty written in standard form? (Lesson 1.2)

(A) 731,934

(B) 731,934,003

(C) 731,934,030

(D) 731,934,300

4. A city has a population of 743,182 people. What is the value of the digit 3? (Lesson 1.2)

(A) 3 hundreds

(B) 3 thousands

(C) 3 ten thousands

(D) 3 thousandths

5. Which expression matches the words "three times the sum of 8 and 4"? (Lesson 1.10)

(A) $3 \times (8 + 4)$

(B) $3 \times 8 + 4$

(C) $3 + 8 \times 4$

(D) $3 \times (8 \times 4)$

6. A family of 2 adults and 3 children goes to a play. Admission costs $8 per adult and $5 per child. Which expression does NOT show the total admission cost for the family?

(Lesson 1.12)

(A) ($8 × 2) + ($5 × 3)

(B) $16 + $15

(C) ($8 × $5) + (2 + 3)

(D) $31

Name _____

Place Value of Decimals

Write the value of the underlined digit.

1. 0.2<u>8</u>7

    8 hundredths, or 0.08

2. 5.<u>3</u>49

_____

3. 2.70<u>4</u>

_____

4. 9.<u>1</u>54

_____

5. 4.00<u>6</u>

_____

6. 7.2<u>5</u>8

_____

7. 0.1<u>9</u>8

_____

8. 6.<u>8</u>21

_____

9. 8.02<u>7</u>

_____

Write the number in two other forms.

10. 0.326

_____

_____

_____

_____

11. 8.517

_____

_____

_____

_____

12. 0.924

_____

_____

_____

13. 1.075

_____

_____

_____

Problem Solving REAL WORLD

14. In a gymnastics competition, Paige's score was 37.025. What is Paige's score written in word form?

_____

15. Jake's batting average for the softball season is 0.368. What is Jake's batting average written in expanded form?

_____

Lesson Check

1. When Mindy went to China, she exchanged $1 for 6.589 Yuan. What digit is in the hundredths place of 6.589?

   (A) 5

   (B) 6

   (C) 8

   (D) 9

2. The diameter of the head of a screw is 0.306 inch. What is this number written in word form?

   (A) three hundred six

   (B) three hundred six thousandths

   (C) thirty-six thousandths

   (D) three and six thousandths

Spiral Review

3. Each car on a commuter train can seat 114 passengers. If the train has 7 cars, how many passengers can the train seat? (Lesson 1.6)

   (A) 770

   (B) 774

   (C) 778

   (D) 798

4. Which of the following expressions has a value of 10? (Lesson 1.11)

   (A) $(9 + 15) \div 3 + 2$

   (B) $9 + (15 \div 3) + 2$

   (C) $9 + 15 \div (3 + 2)$

   (D) $(9 + 15 \div 3) + 2$

5. Danica has 15 stickers. She gives 3 to one friend and gets 4 from another friend. Which expression matches the words? (Lesson 1.10)

   (A) $15 + 3 + 4$

   (B) $15 - (3 + 4)$

   (C) $15 - 3 + 4$

   (D) $15 + 3 - 4$

6. There are 138 people seated at the tables in a banquet hall. Each table can seat 12 people. All the tables are full except one. How many full tables are there? (Lesson 2.7)

   (A) 6

   (B) 11

   (C) 12

   (D) 13

Name _____

Compare and Order Decimals

Compare. Write <, >, or =.

1. 4.735 (<) 4.74

2. 2.549 ◯ 2.549

3. 3.207 ◯ 3.027

4. 8.25 ◯ 8.250

5. 5.871 ◯ 5.781

6. 9.36 ◯ 9.359

7. 1.538 ◯ 1.54

8. 7.036 ◯ 7.035

9. 6.700 ◯ 6.7

Order from greatest to least.

10. 3.008; 3.825; 3.09; 3.18

11. 0.275; 0.2; 0.572; 0.725

_____

_____

12. 6.318; 6.32; 6.230; 6.108

13. 0.456; 1.345; 0.645; 0.654

_____

_____

Algebra Find the unknown digit to make each statement true.

14. 2.48 > 2.4 ▨ 1 > 2.463

15. 5.723 < 5.72 ▨ < 5.725

16. 7.64 < 7. ▨ 5 < 7.68

_____

_____

_____

Problem Solving  REAL WORLD

17. The completion times for three runners in a 100-yard dash are 9.75 seconds, 9.7 seconds, and 9.675 seconds. Which is the winning time?

18. In a discus competition, an athlete threw the discus 63.37 meters, 62.95 meters, and 63.7 meters. Order the distances from least to greatest.

_____

_____

Lesson Check

Jay, Alana, Evan, and Stacey work together to complete a science experiment. The table at the right shows the amount of liquid left in each of their beakers at the end of the experiment.

Student	Amount of liquid (liters)
Jay	0.8
Alana	1.05
Evan	1.2
Stacey	0.75

1. Whose beaker has the greatest amount of liquid left in it?

   (A) Jay   (C) Evan

   (B) Alana   (D) Stacey

2. Whose beaker has the least amount of liquid left in it?

   (A) Jay   (C) Evan

   (B) Alana   (D) Stacey

Spiral Review

3. Janet walked 3.75 miles yesterday. Which is the word form of 3.75? (Lesson 3.2)

   (A) three and seventy-five tenths

   (B) three hundred seventy-five hundredths

   (C) three hundred seventy-five thousandths

   (D) three and seventy-five hundredths

4. A dance school allows a maximum of 15 students per class. If 112 students sign up for dance class, how many classes does the school need to offer to accommodate all the students? (Lesson 2.7)

   (A) 7   (C) 9

   (B) 8   (D) 10

5. Which expression has a value of 7?
   (Lesson 1.12)

   (A) $[(29 - 18) + (17 + 8)] \div 6$

   (B) $[(29 - 18) + (17 - 8)] \div 4$

   (C) $[(29 + 18) - (17 + 8)] \div 2$

   (D) $[(29 + 18) + (17 - 8)] \div 8$

6. Cathy cut 2 apples into 6 slices each. She ate 9 slices. Which expression matches the words? (Lesson 1.10)

   (A) $(2 \times 6) - 9$

   (B) $(6 \times 9) - 2$

   (C) $(9 \times 2) - 6$

   (D) $(9 - 6) \times 2$

Round Decimals

Write the place value of the underlined digit. Round each number to the place of the underlined digit.

1. 0.<u>7</u>82

    _____tenths_____

    _____0.8_____

2. <u>4</u>.735

    _____

    _____

3. 2.<u>3</u>48

    _____

    _____

4. 0.5<u>0</u>6

    _____

    _____

5. 15.<u>1</u>86

    _____

    _____

6. 8.4<u>6</u>5

    _____

    _____

Name the place value to which each number was rounded.

7. 0.546 to 0.55

    _____

8. 4.805 to 4.8

    _____

9. 6.493 to 6

    _____

10. 1.974 to 2.0

    _____

11. 7.709 to 8

    _____

12. 14.637 to 15

    _____

Round 7.954 to the place named.

13. tenths

    _____

14. hundredths

    _____

15. ones

    _____

Round 18.194 to the place named.

16. tenths

    _____

17. hundredths

    _____

18. ones

    _____

Problem Solving  REAL WORLD

19. The population density of Montana is 6.699 people per square mile. What is the population density per square mile of Montana rounded to the nearest whole number?

    _____

20. Alex's batting average is 0.346. What is his batting average rounded to the nearest hundredth?

    _____

Lesson Check

1. Ms. Ari buys and sells diamonds. She has a diamond that weighs 1.825 carats. What is the weight of Ms. Ari's diamond rounded to the nearest hundredth?

   (A) 1.8 carats

   (B) 1.82 carats

   (C) 1.83 carats

   (D) 1.9 carats

2. A machinist uses a special tool to measure the diameter of a small pipe. The measurement tool reads 0.276 inch. What is this measure rounded to the nearest tenth?

   (A) 0.2 inch

   (B) 0.27 inch

   (C) 0.28 inch

   (D) 0.3 inch

Spiral Review

3. Four ice skaters participate in an ice skating competition. The table shows their scores. Who has the highest score? (Lesson 3.3)

Name	Points
Natasha	75.03
Taylor	75.39
Rowena	74.98
Suki	75.3

   (A) Natasha   (C) Rowena

   (B) Taylor   (D) Suki

4. Which of the following statements is true about the relationship between the decimals 0.09 and 0.9? (Lesson 3.1)

   (A) 0.09 is equal to 0.9.

   (B) 0.09 is 10 times as much as 0.9.

   (C) 0.9 is $\frac{1}{10}$ of 0.09.

   (D) 0.09 is $\frac{1}{10}$ of 0.9

5. The population of Foxville is about 12×10^3 people. Which is another way to write this number? (Lesson 1.5)

   (A)   120

   (B)   1,200

   (C)   12,000

   (D) 120,000

6. Joseph needs to find the quotient of 3,216 ÷ 8. In which place is the first digit in the quotient? (Lesson 2.1)

   (A) ones

   (B) tens

   (C) hundreds

   (D) thousands

Name _____

Decimal Addition

Add. Draw a quick picture.

1. $0.5 + 0.6 =$ __**1.1**__

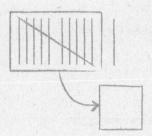

2. $0.15 + 0.36 =$ _____

3. $0.8 + 0.7 =$ _____

4. $0.35 + 0.64 =$ _____

5. $0.54 + 0.12 =$ _____

6. $0.51 + 0.28 =$ _____

7. $3.8 + 1.4 =$ _____

8. $2.71 + 2.15 =$ _____

9. $2.9 + 1.4 =$ _____

Problem Solving REAL WORLD

10. Draco bought 0.6 pound of bananas and 0.9 pound of grapes at the farmers' market. What is the total weight of the fruit?

11. Nancy biked 2.65 miles in the morning and 3.19 miles in the afternoon. What total distance did she bike?

_____

_____

Lesson Check

1. What is the sum of 2.5 and 1.9?

   (A) 0.6

   (B) 1.6

   (C) 3.4

   (D) 4.4

2. Keisha walked 0.65 hour in the morning and 0.31 hour in the evening. How many hours did she walk altogether?

   (A) 0.96 hour

   (B) 0.86 hour

   (C) 0.34 hour

   (D) 0.33 hour

Spiral Review

3. Jodi walks 35 minutes a day. If she walks for 240 days, how many minutes altogether does Jodi walk? (Lesson 1.7)

   (A)   840 minutes

   (B)   850 minutes

   (C) 8,400 minutes

   (D) 8,500 minutes

4. The Speeders soccer team charged $12 to wash each car at a fundraiser car wash. The team collected a total of $672 by the end of the day. How many cars did the team wash? (Lesson 2.6)

   (A) 56

   (B) 57

   (C) 58

   (D) 59

5. David records the number of visitors to the snake exhibit each day for 6 days. His data are shown in the table. If admission is $7 per person, how much money did the snake exhibit make altogether over the 6 days?

   (Lesson 1.6)

Visitors to the Snake Exhibit					
30	25	44	12	25	32

   (A) $42   (C) $308

   (B) $210   (D) $1,176

6. What is the value of the expression? (Lesson 1.11)

   $$6 + 18 \div 3 \times 4$$

   (A) 2

   (B) 30

   (C) 32

   (D) 48

Decimal Subtraction

Subtract. Draw a quick picture.

1. $0.7 - 0.2 =$ __**0.5**__

2. $0.45 - 0.24 =$ _____

3. $0.92 - 0.51 =$ _____

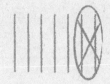

4. $0.67 - 0.42 =$ _____

5. $0.9 - 0.2 =$ _____

6. $3.25 - 1.67 =$ _____

7. $4.1 - 2.7 =$ _____

8. $3.12 - 2.52 =$ _____

9. $3.6 - 1.8 =$ _____

Problem Solving REAL WORLD

10. Yelina made a training plan to run 5.6 miles per day. So far, she has run 3.1 miles today. How much farther does she have to run to meet her goal for today?

11. Tim cut a 2.3-foot length of pipe from a pipe that was 4.1 feet long. How long is the remaining piece of pipe?

_____

_____

Lesson Check

1. Janice wants to jog 3.25 miles on the treadmill. She has jogged 1.63 miles. How much farther does she have to jog to meet her goal?

   (A) 1.68 miles

   (B) 1.62 miles

   (C) 1.58 miles

   (D) 1.52 miles

2. A new teen magazine has a readership goal of 3.5 million. Its current readership is 2.8 million. How much does its readership need to increase to meet this goal?

   (A) 0.7 million

   (B) 1.7 million

   (C) 5.3 million

   (D) 6.3 million

Spiral Review

3. What is the value of the underlined digit in 91,764,350? **(Lesson 1.2)**

   (A) 700,000

   (B) 70,000

   (C) 7,000

   (D) 700

4. How many zeros are in the product $(6 \times 5) \times 10^3$? **(Lesson 1.5)**

   (A) 3

   (B) 4

   (C) 5

   (D) 6

5. To evaluate the following expression, which step should you do first? **(Lesson 1.12)**

   $7 \times (4 + 16) \div 4 - 2$

   (A) Multiply 7 and 4.

   (B) Add 4 and 16.

   (C) Divide 16 by 4.

   (D) Subtract 2 from 4.

6. In the past two weeks, Sue earned $513 at her part-time job. She worked a total of 54 hours. About how much did Sue earn per hour? **(Lesson 2.5)**

   (A) $5

   (B) $10

   (C) $12

   (D) $15

Estimate Decimal Sums and Differences

Use rounding to estimate.

1.   5.38
    +6.14

$$\begin{array}{r} 5 \\ +6 \\ \hline 11 \end{array}$$

2.   2.57
    +0.14

_____

3.   9.65
    −3.12

_____

4.   7.92
    +5.37

_____

Use benchmarks to estimate.

5.   2.81
    +3.72

_____

6.   12.54
    + 7.98

_____

7.   6.34
    +3.95

_____

8.   16.18
    − 5.94

_____

9.   17.09
    + 3.98

_____

10.   14.01
    − 4.51

_____

11.   11.47
    + 9.02

_____

12.   19.97
    −11.02

_____

Problem Solving  REAL WORLD

13. Elian bought 1.87 pounds of chicken and 2.46 pounds of turkey at the deli. About how much meat did he buy altogether?

_____

14. Jenna bought a gallon of milk at the store for $3.58. About how much change did she receive from a $20 bill?

_____

Lesson Check

1. Regina has two electronic files. One has a size of 3.15 MB and the other has a size of 4.89 MB. Which is the best estimate of the total size of the two electronic files?

   (A) 7 MB

   (B) 7.5 MB

   (C) 8 MB

   (D) 8.5 MB

2. Madison is training for a marathon, which is 26.2 miles. She currently can run 18.5 miles in a day. About how many more miles in a day does she have to add to run the length of a marathon?

   (A) 8 miles

   (B) 7.5 miles

   (C) 6.5 miles

   (D) 6 miles

Spiral Review

3. A machine prints 8 banners in 120 seconds. How many seconds does it take to print one banner? (Lesson 2.2)

   (A) 10 seconds

   (B) 12 seconds

   (C) 15 seconds

   (D) 18 seconds

4. To which place value is the number rounded? (Lesson 3.4)

   5.319 to 5.3

   (A) ones

   (B) tenths

   (C) hundredths

   (D) thousandths

5. The average distance from Mars to the sun is about one hundred forty-one million, six hundred twenty thousand miles. How do you write the number that shows this distance in standard form? (Lesson 1.2)

   (A)   141,620

   (B)   1,416,200

   (C)   14,162,000

   (D)   141,620,000

6. Logan ate 1.438 pounds of grapes. His brother Ralph ate 1.44 pounds of grapes. Which brother ate more grapes? (Lesson 3.3)

   (A) Logan

   (B) Ralph

   (C) They ate the same amount of grapes.

   (D) There is not enough information to decide which brother ate more grapes.

Name _____

Add Decimals

Estimate. Then find the sum.

1. Estimate: _____

    2.85
  +7.29

   1 1
  2.85
 +7.29
 10.14

2. Estimate: _____

    4.23
  +6.51

3. Estimate: _____

    6.8
  +4.2

4. Estimate: _____

    2.7
  +5.37

Find the sum.

5. 6.8 + 4.4

6. 6.87 + 5.18

7. 3.14 + 2.9

8. 16.18 + 5.94

9. 19.8 + 31.45

10. 25.47 + 7.24

11. 9.17 + 5.67

12. 19.7 + 5.46

Problem Solving  REAL WORLD

13. Marcela's dog gained 4.1 kilograms in two months. Two months ago, the dog's mass was 5.6 kilograms. What is the dog's current mass?

14. During last week's storm, 2.15 inches of rain fell on Monday and 1.68 inches of rain fell on Tuesday. What was the total amount of rainfall on both days?

_____

_____

Lesson Check

1. Lindsay has two packages she wants to mail. One package weighs 6.3 ounces, and the other package weighs 4.9 ounces. How much do the packages weigh together?

- (A) 11.4 ounces
- (B) 11.2 ounces
- (C) 10.9 ounces
- (D) 10.5 ounces

2. Anton rode his mountain bike three days in a row. He biked 12.1 miles on the first day, 13.4 miles on the second day, and 17.9 miles on the third day. How many total miles did Anton bike during the three days?

- (A) 58.2 miles
- (B) 47.1 miles
- (C) 43.4 miles
- (D) 42.4 miles

Spiral Review

3. In the number 2,145,857, how does the digit 5 in the thousands place compare to the digit 5 in the tens place? **(Lesson 1.1)**

- (A) It is 10 times greater.
- (B) It is 100 times greater.
- (C) It is 1,000 times greater.
- (D) It is 10,000 times greater.

4. Which of the following expressions does NOT have the same value as 10^5? **(Lesson 1.4)**

- (A) $10 \times 10 \times 10 \times 10 \times 10$
- (B) 100,000
- (C) the fifth power of 10
- (D) $5 \times 10,000$

5. Carmen works at a pet store. To feed 8 cats, she empties four 6-ounce cans of cat food into a large bowl. Carmen divides the food equally among the cats. How many ounces of food will each cat get? **(Lesson 1.9)**

- (A) 3 ounces
- (B) 4 ounces
- (C) 6 ounces
- (D) 8 ounces

6. There are 112 students in the Hammond Middle School marching band. The band director wants the students to march with 14 students in each row for the upcoming parade. How many rows will there be?

(Lesson 2.3)

- (A) 8
- (B) 10
- (C) 12
- (D) 14

Name _____

Subtract Decimals

Estimate. Then find the difference.

1. Estimate: __**3**__

   6.5
   −3.9
   _____

   $\overset{5\ 15}{\cancel{6}.\cancel{5}}$
   −3.9
   _____
   2.6
   _____

2. Estimate: _____

   4.23
   −2.51
   _____

   _____

3. Estimate: _____

   8.6
   −5.1
   _____

   _____

4. Estimate: _____

   2.71
   −1.34
   _____

   _____

Find the difference. Check your answer.

5.   16.3
    − 4.4
   _____

   _____

6.   12.56
    − 5.18
   _____

   _____

7.   3.14
    −2.9
   _____

   _____

8.   34.9
    − 4.29
   _____

   _____

9. $2.54 - 1.67$

   _____

10. $25.8 - 14.7$

   _____

11. $11.63 - 6.7$

   _____

12. $5.24 - 2.14$

   _____

Problem Solving REAL WORLD

13. The width of a tree was 3.15 inches last year. This year, the width is 5.38 inches. How much did the width of the tree increase?

   _____

14. The temperature decreased from 71.5°F to 56.8°F overnight. How much did the temperature drop?

   _____

Lesson Check

1. During training, Janice kayaked 4.68 miles on Monday and 5.61 miles on Tuesday. How much farther did she kayak on Tuesday?

   (A) 1.13 miles

   (B) 1.03 miles

   (C) 0.93 mile

   (D) 0.83 mile

2. Devon had a length of rope that was 4.78 meters long. He cut a 1.45-meter length from it. How much rope does he have left?

   (A) 6.23 meters

   (B) 5.13 meters

   (C) 3.33 meters

   (D) 2.33 meters

Spiral Review

3. A dairy farm has 9 pastures and 630 cows. The same number of cows are placed in each pasture. How many cows are in each pasture? (Lesson 2.2)

   (A) 60

   (B) 70

   (C) 600

   (D) 700

4. Moya records 6.75 minutes of an interview on one tape and 3.75 minutes of the interview on another tape. How long was the total interview? (Lesson 3.8)

   (A) 9.25 minutes

   (B) 9.5 minutes

   (C) 10.25 minutes

   (D) 10.5 minutes

5. Joanna, Dana, and Tracy shared some trail mix. Joanna ate 0.125 pound of trail mix, Dana ate 0.1 pound, and Tracy ate 0.12 pound of trail mix. Which lists the friends in order from least to greatest amount of trail mix eaten? (Lesson 3.3)

   (A) Dana, Tracy, Joanna

   (B) Joanna, Tracy, Dana

   (C) Tracy, Dana, Joanna

   (D) Joanna, Dana, Tracy

6. The local park has 4 bike racks. Each bike rack can hold 15 bikes. There are 16 bikes in the bike racks. Which expression shows the total number of empty spaces in the bike racks? (Lesson 1.11)

   (A) $(15 \times 16) + 4$

   (B) $(15 \times 16) - 4$

   (C) $(4 \times 15) + 16$

   (D) $(15 \times 4) - 16$

Name _____

Patterns with Decimals

Write a rule for the sequence. Then find the unknown term.

1. 2.6, 3.92, 5.24, **6.56**, 7.88

   Think: 2.6 + ? = 3.92; 3.92 + ? = 5.24

   2.6 + 1.32 = 3.92
   3.92 + 1.32 = 5.24

   Rule: _____ **add 1.32** _____

2. 25.7, 24.1, _____, 20.9, 19.3

   Rule: _____

3. 14.33, 13.22, 12.11, 11.00, _____

   Rule: _____

4. 1.75, _____, 6.75, 9.25, 11.75

   Rule: _____

Write the first four terms of the sequence.

5. **Rule:** start at 17.3, add 0.9

6. **Rule:** start at 28.6, subtract 3.1

_____, _____, _____, _____

_____, _____, _____, _____

Problem Solving REAL WORLD

7. The Ride-It Store rents bicycles. The cost is $8.50 for 1 hour, $13.65 for 2 hours, $18.80 for 3 hours, and $23.95 for 4 hours. If the pattern continues, how much will it cost Nate to rent a bike for 6 hours?

_____

8. Lynne walks dogs every day to earn money. The fees she charges per month are 1 dog, $40; 2 dogs, $37.25 each; 3 dogs, $34.50 each; 4 dogs, $31.75 each. A pet store wants her to walk 8 dogs. If the pattern continues, how much will Lynne charge to walk each of the 8 dogs?

_____

Lesson Check

1. A store has a sale on books. The price is $17.55 for one book, $16.70 each for 2 books, $15.85 each for 3 books, and $15 each for 4 books. If this pattern continues, how much will it cost to buy 7 books?

- (A) $14.15 each
- (B) $13.30 each
- (C) $13.15 each
- (D) $12.45 each

2. A bowling alley offers special weekly bowling rates. The weekly rates are 5 games for $15, 6 games for $17.55, 7 games for $20.10, and 8 games for $22.65. If this pattern continues, how much will it cost to bowl 10 games in a week?

- (A) $25.20
- (B) $27.75
- (C) $28.20
- (D) $37.95

Spiral Review

3. Find the product. (Lesson 1.7)

$$284 \times 36$$

- (A) 2,556
- (B) 7,704
- (C) 9,224
- (D) 10,224

4. At a sale, a shoe store sold 8 pairs of shoes for a total of $256. Each pair cost the same amount. What was the price of each pair of shoes? (Lesson 2.2)

- (A) $22
- (B) $32
- (C) $248
- (D) $2,048

5. Marcie jogged 0.8 mile on Wednesday and 0.9 mile on Thursday. How far did she jog altogether? (Lesson 3.8)

- (A) 0.1 mile
- (B) 0.17 mile
- (C) 1.1 miles
- (D) 1.7 miles

6. Bob has 5.5 cups of flour. He uses 3.75 cups of flour. How much flour does Bob have left? (Lesson 3.9)

- (A) 2.75 cups
- (B) 2.25 cups
- (C) 1.75 cups
- (D) 1.25 cups

Name _____

Problem Solving • Add and Subtract Money

Solve. Use the table to solve 1–3.

1. Dorian and Jack decided to go bowling. They each
   need to rent shoes and 1 lane, and Jack is a member.
   If Jack pays for both of them with $20, what change should
   he receive?

Bowl-a-Rama		
	Regular Cost	Member's Cost
Lane Rental (up to 4 people)	$9.75	$7.50
Shoe Rental	$3.95	$2.95

   Calculate the cost: $7.50 + $3.95 + $2.95 = $14.40

   Calculate the change: $20 − $14.40 = $5.60

2. Natalie and her friends decided to rent 4 lanes at regular
   cost for a party. Ten people need to rent shoes, and 4
   people are members. What is the total cost for the party?

   _____

3. Warren paid $23.85 and received no change. He is a
   member and rented 2 lanes. How many pairs of
   shoes did he rent?

   _____

Use the following information to solve 4–6.

At the concession stand, medium sodas cost $1.25 and
hot dogs cost $2.50.

4. Natalie's group brought in pizzas, but is buying the drinks at
   the concession stand. How many medium sodas can Natalie's
   group buy with $20? Make a table to show your answer.

   _____

5. Jack bought 2 medium sodas and 2 hot dogs. He paid
   with $20. What was his change?

   _____

6. How much would it cost to buy 3 medium sodas and 2 hot dogs?

   _____

© Houghton Mifflin Harcourt Publishing Company

Lesson Check

1. Prakrit bought a pack of paper for $5.69 and printer toner for $9.76. He paid with a $20 bill. What was his change?

   (A) $5.55
   (B) $5.45
   (C) $4.55
   (D) $4.45

2. Elysse paid for her lunch with a $10 bill and received $0.63 in change. The lunch special was $7.75. Sales tax was $0.47. What was the cost of her drink?

   (A) $1.15
   (B) $1.97
   (C) $2.87
   (D) $2.97

Spiral Review

3. Tracie has saved $425 to spend during her 14-day vacation. About how much money can she spend each day? (Lesson 2.5)

   (A) $45
   (B) $42
   (C) $30
   (D) $14

4. Which of the following decimals is $\frac{1}{10}$ of 0.08? (Lesson 3.1)

   (A) 8.0
   (B) 0.8
   (C) 0.18
   (D) 0.008

5. Tyrone bought 2.25 pounds of Swiss cheese and 4.2 pounds of turkey at the deli. About how much was the weight of the two items? (Lesson 3.7)

   (A) 6 pounds
   (B) 7 pounds
   (C) 8 pounds
   (D) 29 pounds

6. Shelly ate 4.2 ounces of trail mix. Marshall ate 4.25 ounces of trail mix. How much more trail mix did Marshall eat? (Lesson 3.9)

   (A) 0.45 ounce
   (B) 0.27 ounce
   (C) 0.23 ounce
   (D) 0.05 ounce

Name _____

Choose a Method

Find the sum or difference.

1. 7.24
    +3.18

    1
    7.24
    +3.18
    10.42

2. 5.2
    6.47
    +12.16

3. 6.37
    −4.98

4. 0.64
    9.68
    +1.47

5. 14.87
    +3.65

6. 60.12
    −14.05

7. 2.72
    +9.48

8. 16.85
    +83.4

9. $13.60 − $8.74 _____

10. $25.00 − $16.32 _____

11. 13.65 + 6.90 + 4.35 _____

Problem Solving REAL WORLD

12. Jill bought 6.5 meters of blue lace and 4.12 meters of green lace. What was the total length of lace she bought?

13. Zack bought a coat for $69.78. He paid with a $100 bill and received $26.73 in change. How much was the sales tax?

_____   _____

Lesson Check

1. Jin buys 4 balls of yarn for a total of $23.78. She pays with two $20 bills. What is her change?

   (A) $1.78

   (B) $3.78

   (C) $16.22

   (D) $18.22

2. Allan is measuring his dining room table to make a tablecloth. The table is 0.45 meter longer than it is wide. If it is 1.06 meters wide, how long is it?

   (A) 1.51 meters

   (B) 1.41 meters

   (C) 1.01 meters

   (D) 1.10 meters

Spiral Review

3. Which of the following can be used to find $56 \div 4$? (Lesson 1.8)

   (A) $(4 \times 7) + (4 \times 8)$

   (B) $(4 \times 50) + (4 \times 6)$

   (C) $(2 \times 28) + (2 \times 2)$

   (D) $(4 \times 10) + (4 \times 4)$

4. Jane, Andre, and Maria pick apples. Andre picks three times as many pounds as Maria. Jane picks two times as many pounds as Andre. The total weight of the apples is 840 pounds. How many pounds of apples does Andre pick? (Lesson 2.9)

   (A) 84 pounds

   (B) 252 pounds

   (C) 504 pounds

   (D) 840 pounds

5. What is the sum of $6.43 + 0.89$? (Lesson 3.8)

   (A) 5.54

   (B) 6.22

   (C) 6.32

   (D) 7.32

6. Hannah bought a total of 5.12 pounds of fruit at the market. She bought 2.5 pounds of pears, and she also bought some bananas. How many pounds of bananas did she buy? (Lesson 3.9)

   (A) 2.37 pounds

   (B) 2.62 pounds

   (C) 3.37 pounds

   (D) 3.5 pounds

Name _____

Chapter 3 Extra Practice

Lessons 3.1 - 3.2

Complete the sentence.

1. 0.7 is 10 times as much as _____ .

2. 0.003 is $\frac{1}{10}$ of _____ .

Write the value of the underlined digit.

3. 3.8̲72

4. 0.194̲

5. 11.77̲6

6. 4.001̲

_____   _____   _____   _____

Lessons 3.3 - 3.4

Order from greatest to least.

1. 5.006, 5.917, 5.08, 5.99

2. 0.823, 1.823, 0.732, 0.832

_____   _____

Write the place value of the underlined digit. Round each number to the place of the underlined digit.

3. 0.8̲29

4. 7̲.918

5. 11.50̲7

_____   _____   _____

Lessons 3.5 - 3.9

Estimate. Then find the sum or difference.

1. Estimate: _____

$$\begin{array}{r} 8.5 \\ + 1.8 \\ \hline \end{array}$$

2. Estimate: _____

$$\begin{array}{r} 26.42 \\ - 9.8 \\ \hline \end{array}$$

3. Estimate: _____

$$\begin{array}{r} 8.26 \\ + 0.47 \\ \hline \end{array}$$

_____   _____   _____

4. Estimate: _____

7.06 − 1.95

5. Estimate: _____

24 − 5.392

6. Estimate: _____

3.6 + 2.16 + 1.34

_____   _____   _____

Lesson 3.10

Write a rule for the sequence. Then, find the unknown term.

1. 56.38, 51.28, 46.18, 41.08, _____

   Rule: _____

2. 2.1, 4.3, 6.5, _____, 10.9

   Rule: _____

3. 15.24, 15.14, 15.04, _____, 14.84

   Rule: _____

4. 9.98, 13.09, 16.2, _____, 22.42

   Rule: _____

Lesson 3.11

Complete the table to solve.

1. Alicia's goal is to have $30 in her account. She starts with $24.50 from washing cars. She spends $8.25 at the movies. Then she earns another $11.50. Does Alicia now have $30? By how much is she over or under?

Alicia's Account	
Starting Balance	$24.50
Spent: $8.25	_____
	_____
Earned: $11.50	_____
Current Balance	_____

_____

Lesson 3.12

Find the sum or difference.

1. 4.15 + 3.55 + 1.85

2. $25 − $12.35

3. 2.74
   +9.36

4. 12.15
   − 6.13

_____   _____   _____   _____

School-Home Letter

Vocabulary

decimal A number with one or more digits to the right of the decimal point

expanded form A way to write numbers by showing the value of each digit

product The answer to a multiplication problem

Dear Family,

Throughout the next few weeks, our math class will be learning about decimal multiplication. We will also be learning how to estimate decimal products.

You can expect to see homework that involves multiplication of decimals.

Here is a sample of how your child will be taught to multiply decimals.

🔑 MODEL Multiply Decimals

Multiply. 3.2×4.17

STEP 1	STEP 2	STEP 3
Estimate.	Multiply as with whole numbers.	Use the estimate to place the decimal point.
3.2×4.17 ↓ ↓ $3 \times 4 = 12$	$\begin{array}{r} 2 \\ 1 \\ 417 \\ \times\ 32 \\ \hline 834 \\ +12{,}510 \\ \hline 13{,}344 \end{array}$	$3.2 \times 4.17 = 13.344$ Think: The product should be close to the estimate.

Tips

Placing the Decimal Point

To help place the decimal point in the product, add the number of decimal places in each factor.

For example, since 4.17 has 2 decimal places and 3.2 has 1 decimal place, the product will have 2 + 1, or 3 decimal places.

Activity

A trip to the grocery store or the gas station is a perfect opportunity to practice decimal operations. For example, "We bought 8.6 gallons of gasoline that cost $2.95 per gallon. What was the total cost?" Work together to write a multiplication sentence with decimals that represents the situation. Then estimate before multiplying to find the exact product.

Carta
para la casa

Vocabulario

decimal Un número con uno o más dígitos a la derecha del punto decimal

forma desarrollada Una manera de escribir los números mostrando el valor de cada dígito

producto El resultado de una multiplicación

Querida familia,

Durante las próximas semanas, aprenderemos acerca de la multiplicación con números decimales. También aprenderemos a estimar productos decimales.

Llevaré a la casa tareas con multiplicaciones de decimales.

Este es un ejemplo de cómo vamos a multiplicar decimales.

🔑 MODELO Multiplicar decimales

Multiplica. 3.2×4.17

PASO 1

Estima.

3.2×4.17

↓   ↓

$3 \times 4 = 12$

PASO 2

Multiplica igual que con los números enteros.

$$\begin{array}{r} 2 \\ 1 \\ 417 \\ \times\ 32 \\ \hline 834 \\ +12{,}510 \\ \hline 13{,}344 \end{array}$$

PASO 3

Usa la estimación para poner el punto decimal.

$3.2 \times 4.17 = 13.344$

Piensa: El producto debe estar cerca de la estimación.

Pistas

Poner el punto decimal

Para poner el punto decimal en el producto, suma el número de lugares decimales en cada factor.

Por ejemplo, como 4.17 tiene 2 lugares decimales y 3.2 tiene 1, el producto tendrá $2 + 1$, o 3 lugares decimales.

Actividad

Una visita a la tienda o a la gasolinera es una buena oportunidad para practicar operaciones con números decimales. Por ejemplo: "Compramos 8.6 galones de gasolina a $2.95 por galón. ¿Cuál fue el precio total?" Trabajen juntos para escribir un enunciado de multiplicación con decimales que represente el evento. Después, estimen antes de multiplicar para hallar el producto exacto.

Name _____

Multiplication Patterns with Decimals

Complete the pattern.

1. $2.07 \times 1 =$ __**2.07**__

$2.07 \times 10 =$ __**20.7**__

$2.07 \times 100 =$ __**207**__

$2.07 \times 1{,}000 =$ __**2,070**__

2. $1 \times 30 =$ _____

$0.1 \times 30 =$ _____

$0.01 \times 30 =$ _____

3. $10^0 \times 0.23 =$ _____

$10^1 \times 0.23 =$ _____

$10^2 \times 0.23 =$ _____

$10^3 \times 0.23 =$ _____

4. $390 \times 1 =$ _____

$390 \times 0.1 =$ _____

$390 \times 0.01 =$ _____

5. $10^0 \times 49.32 =$ _____

$10^1 \times 49.32 =$ _____

$10^2 \times 49.32 =$ _____

$10^3 \times 49.32 =$ _____

6. $1 \times 9{,}670 =$ _____

$0.1 \times 9{,}670 =$ _____

$0.01 \times 9{,}670 =$ _____

7. $874 \times 1 =$ _____

$874 \times 10 =$ _____

$874 \times 100 =$ _____

$874 \times 1{,}000 =$ _____

8. $10^0 \times 10 =$ _____

$10^1 \times 10 =$ _____

$10^2 \times 10 =$ _____

$10^3 \times 10 =$ _____

9. $1 \times 5 =$ _____

$0.1 \times 5 =$ _____

$0.01 \times 5 =$ _____

Problem Solving REAL WORLD

10. Nathan plants equal-sized squares of sod in his front yard. Each square has an area of 6 square feet. Nathan plants a total of 1,000 squares in his yard. What is the total area of the squares of sod?

11. Three friends are selling items at a bake sale. May makes $23.25 selling bread. Inez sells gift baskets and makes 100 times as much as May. Carolyn sells pies and makes one tenth of the money Inez makes. How much money does each friend make?

_____

_____

_____

_____

Lesson Check

1. The length of the Titanic was 882 feet. Porter's history class is building a model of the Titanic. The model is $\frac{1}{100}$ of the actual length of the ship. How long is the model?

   (A) 882 feet

   (B) 88.2 feet

   (C) 8.82 feet

   (D) 0.882 feet

2. Ted is asked to multiply $10^2 \times 18.72$. How should he move the decimal point to get the correct product?

   (A) 2 places to the right

   (B) 1 place to the right

   (C) 1 place to the left

   (D) 2 places to the left

Spiral Review

3. The table shows the height in meters of some of the world's tallest buildings. Which list shows the heights in order from least to greatest? (Lesson 3.3)

Building	Height (meters)
Zifeng Tower	457.2
International Finance Center	415.138
Burj Khalifa	828.142
Petronas Towers	452.018

   (A) 457.2, 415.138, 828.142, 452.018

   (B) 415.138, 457.2, 452.018, 828.142

   (C) 828.142, 457.2, 452.018, 415.138

   (D) 415.138, 452.018, 457.2, 828.142

4. Madison had $187.56 in her checking account. She deposited $49.73 and then used her debit card to spend $18.64. What is Madison's new account balance? (Lesson 3.11)

   (A) $119.19

   (B) $218.65

   (C) $237.29

   (D) $255.93

5. What is 3.47 rounded to the nearest tenth? (Lesson 3.4)

   (A) 3.0

   (B) 3.4

   (C) 3.5

   (D) 4.0

6. The city gardener ordered 1,680 tulip bulbs for Riverside Park. The bulbs were shipped in 35 boxes with an equal number of bulbs in each box. How many tulip bulbs were in each box? (Lesson 2.6)

   (A) 47   (C) 57

   (B) 48   (D) 58

Name _____

Multiply Decimals and Whole Numbers

Use the decimal model to find the product.

1. $4 \times 0.07 =$ **0.28**

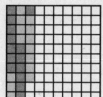

2. $3 \times 0.27 =$ _____

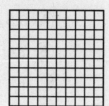

3. $2 \times 0.45 =$ _____

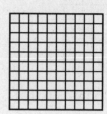

Find the product. Draw a quick picture.

4. $2 \times 0.8 =$ _____

5. $3 \times 0.33 =$ _____

6. $5 \times 0.71 =$ _____

7. $4 \times 0.23 =$ _____

Problem Solving

8. In physical education class, Sonia walks a distance of 0.12 mile in 1 minute. At that rate, how far can she walk in 9 minutes?

9. A certain tree can grow 0.45 meter in one year. At that rate, how much can the tree grow in 3 years?

Lesson Check

1. The model below represents which multiplication sentence?

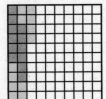

   (A) $6 \times 0.04 = 0.24$

   (B) $4 \times 0.06 = 0.24$

   (C) $8 \times 0.03 = 0.24$

   (D) $3 \times 0.08 = 0.24$

2. A certain type of lunch meat contains 0.5 grams of unsaturated fat per serving. How much unsaturated fat is in 3 servings of the lunch meat?

   (A) 3.5 grams

   (B) 3 grams

   (C) 1.5 grams

   (D) 0.5 gram

Spiral Review

3. To find the value of the following expression, which operation should you do first?
   (Lesson 1.12)

   $$20 - (7 + 4) \times 5$$

   (A) Subtract 7 from 20.

   (B) Add 7 and 4.

   (C) Multiply 4 and 5.

   (D) It does not matter which operation you do first.

4. Ella and three friends run in a relay race that is 14 miles long. Each person runs equal parts of the race. How many miles does each person run? **(Lesson 2.7)**

   (A) 3 miles

   (B) $3\frac{1}{2}$ miles

   (C) 4 miles

   (D) $4\frac{2}{3}$ miles

5. Which statement about 17.518 and 17.581 is true? **(Lesson 3.3)**

   (A) $17.518 < 17.581$

   (B) $17.518 > 17.581$

   (C) $17.518 = 17.581$

   (D) $17.581 < 17.518$

6. Each number in the following sequence has the same relationship to the number immediately before it. How can you find the next number in the sequence? **(Lesson 1.5)**

   $$3, 30, 300, 3{,}000, \ldots$$

   (A) Multiply the previous number by 3.

   (B) Multiply the previous number by 30.

   (C) Multiply the previous number by 10.

   (D) Multiply the previous number by 100.

Name _____

Multiplication with Decimals and Whole Numbers

Find the product.

1.
```
   2.7
×    4
──────
 10.8
```
Think: The place value of the decimal factor is tenths.

2.
```
   7.6
×    8
──────
```

3.
```
  0.35
×    6
──────
```

4.
```
  8.42
×    9
──────
```

5.
```
 14.05
×    7
──────
```

6.
```
 23.82
×    5
──────
```

7. 4×9.3

8. 3×7.9

9. 5×42.89

10. 8×2.6

11. 6×0.92

12. 9×1.04

13. 7×2.18

14. 3×19.54

Problem Solving  REAL WORLD

15. A half-dollar coin issued by the United States Mint measures 30.61 millimeters across. Mikk has 9 half dollars. He lines them up end to end in a row. What is the total length of the row of half dollars?

16. One pound of grapes costs $3.49. Linda buys exactly 3 pounds of grapes. How much will the grapes cost?

_____   _____

Lesson Check

1. Pete wants to make turkey sandwiches for two friends and himself. He wants each sandwich to contain 3.5 ounces of turkey. How many ounces of turkey does he need?

   (A) 3.5 ounces

   (B) 7 ounces

   (C) 10.5 ounces

   (D) 14 ounces

2. Gasoline costs $2.84 per gallon. Mary's father puts 9 gallons of gasoline in the tank of his car. How much will the gasoline cost?

   (A) $2.84

   (B) $9

   (C) $25.56

   (D) $255.60

Spiral Review

3. A group of 5 boys and 8 girls goes to the fair. Admission costs $9 per person. Which expression does NOT show the total amount the group will pay? **(Lesson 1.11)**

   (A) $9 × (5 + 8)

   (B) $9 × 5 × 8

   (C) ($9 × 5) + ($9 × 8)

   (D) $9 × 13

4. Sue and 4 friends buy a box of 362 baseball cards at a yard sale. If they share the cards equally, how many cards will each person receive? **(Lesson 2.2)**

   (A) 91

   (B) 90

   (C) 73

   (D) 72

5. Sarah rides her bicycle 2.7 miles to school. She takes a different route home, which is 2.5 miles. How many miles does Sarah ride to and from school each day? **(Lesson 3.8)**

   (A) 2.5 miles

   (B) 2.7 miles

   (C) 5.2 miles

   (D) 5.4 miles

6. Tim has a box of 15 markers. He gives 3 markers each to 4 friends. Which expression shows the number of markers Tim has left? **(Lesson 1.10)**

   (A) (3 × 4) − 15

   (B) 15 + (3 × 4)

   (C) (15 × 4) − 3

   (D) 15 − (3 × 4)

Name _____

Multiply Using Expanded Form

Draw a model to find the product.

1. $37 \times 9.5 =$ __**351.5**__

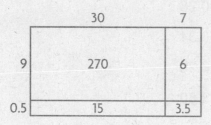

2. $84 \times 0.24 =$ _____

Find the product.

3. $13 \times 0.53 =$ _____   4. $27 \times 89.5 =$ _____   5. $32 \times 12.71 =$ _____

6. $17 \times 0.52 =$ _____   7. $23 \times 59.8 =$ _____   8. $61 \times 15.98 =$ _____

Problem Solving REAL WORLD

9. An object that weighs one pound on the moon will weigh about 6.02 pounds on Earth. Suppose a moon rock weighs 11 pounds on the moon. How much will the same rock weigh on Earth?

10. Tessa is on the track team. For practice and exercise, she runs 2.25 miles each day. At the end of 14 days, how many total miles will Tessa have run?

_____   _____

Lesson Check

1. A baker is going to make 24 blueberry pies. She wants to make sure each pie contains 3.5 cups of blueberries. How many cups of blueberries will she need?

   (A) 3.5 cups

   (B) 6.86 cups

   (C) 24 cups

   (D) 84 cups

2. Aaron buys postcards while he is on vacation. It costs $0.28 to send one postcard. Aaron wants to send 12 postcards. How much will it cost Aaron to send all the postcards?

   (A) $0.28

   (B) $0.34

   (C) $3.36

   (D) $33.60

Spiral Review

3. What is the value of the digit 4 in the number 524,897,123? **(Lesson 1.2)**

   (A)   4,000

   (B)   40,000

   (C)   400,000

   (D) 4,000,000

4. How many zeros will be in the product $(6 \times 5) \times 10^3$? **(Lesson 1.5)**

   (A) 2

   (B) 3

   (C) 4

   (D) 5

5. Roast beef costs $8.49 per pound. What is the cost of 2 pounds of roast beef? **(Lesson 4.3)**

   (A) $8.49

   (B) $10.49

   (C) $16.98

   (D) $169.80

6. North Ridge Middle school collected 5,024 cans of food for a food drive. Each of the 18 homerooms collected about the same number of cans. About how many cans did each homeroom collect? **(Lesson 2.5)**

   (A) 250   (C) 500

   (B) 400   (D) 800

Name _____

Problem Solving • Multiply Money

Solve each problem.

1. Three friends go to the local farmers'
   market. Ashlee spends $8.25. Natalie
   spends 4 times as much as Ashlee. Patrick
   spends $9.50 more than Natalie. How
   much does Patrick spend?

   Ashlee | $8.25

   Natalie | $8.25 | $8.25 | $8.25 | $8.25

   $4 \times \$8.25 = \$33.00$

   Patrick | $8.25 | $8.25 | $8.25 | $8.25 | $9.50

   $\$33.00 + \$9.50 = \$42.50$

   $42.50
   _____

2. Kimmy's savings account has a balance of
   $76.23 in June. By September, her balance is
   5 times as much as her June balance. Between
   September and December, Kimmy deposits a
   total of $87.83 into her account. If she does not
   withdraw any money from her account, what
   should Kimmy's balance be in December?

   _____

3. Amy raises $58.75 to participate in a walk-a-thon.
   Jeremy raises $23.25 more than Amy. Oscar raises
   3 times as much as Jeremy. How much money
   does Oscar raise?

   _____

4. It costs $5.50 per hour to rent a pair of ice skates,
   for up to 2 hours. After 2 hours, the rental cost per
   hour decreases to $2.50. How much does it cost to
   rent a pair of ice skates for 4 hours?

   _____

Lesson Check

1. A family of two adults and four children is going to an amusement park. Admission is $21.75 for adults and $15.25 for children. What is the total cost of the family's admission?

   Ⓐ $37

   Ⓑ $89.25

   Ⓒ $104.50

   Ⓓ $117.50

2. Ms. Rosenbaum buys 5 crates of apples at the market. Each crate costs $12.50. She also buys one crate of pears for $18.75. What is the total cost of the apples and pears?

   Ⓐ $12.50

   Ⓑ $31.25

   Ⓒ $62.50

   Ⓓ $81.25

Spiral Review

3. How do you write $10 \times 10 \times 10 \times 10$ using exponents? **(Lesson 1.4)**

   Ⓐ 10^3

   Ⓑ 10^4

   Ⓒ $10,000$

   Ⓓ 4^{10}

4. Which represents 125.638 rounded to the nearest hundredth? **(Lesson 3.4)**

   Ⓐ 100

   Ⓑ 125.6

   Ⓒ 125.63

   Ⓓ 125.64

5. The sixth-graders at Meadowbrook Middle School are going on a field trip. The 325 students and adults will ride in school buses. Each bus holds 48 people. How many school buses are needed? **(Lesson 2.7)**

   Ⓐ 6

   Ⓑ 6.77

   Ⓒ 7

   Ⓓ 8

6. A restaurant can seat 100 people. It has booths that seat 4 people and tables that seat 6 people. So far, 5 of the booths are full. Which expression matches the situation? **(Lesson 1.10)**

   Ⓐ $4 \times 5 + 6$

   Ⓑ $100 - (5 \times 4)$

   Ⓒ $6 \times (4 + 5)$

   Ⓓ $100 \div (4 + 6)$

Name _____

Decimal Multiplication

Multiply. Use the decimal model.

1. $0.3 \times 0.6 =$ ___0.18___

2. $0.2 \times 0.8 =$ _____

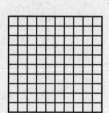

3. $0.5 \times 1.7 =$ _____

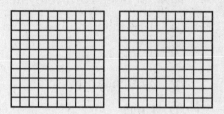

4. $0.6 \times 0.7 =$ _____

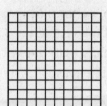

5. $0.8 \times 0.5 =$ _____

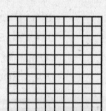

6. $0.4 \times 1.9 =$ _____

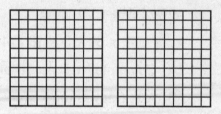

7. $0.8 \times 0.8 =$ _____

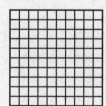

8. $0.2 \times 0.5 =$ _____

9. $0.8 \times 1.3 =$ _____

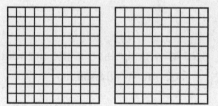

Problem Solving

10. A certain bamboo plants grow 1.2 feet in 1 day. At that rate, how many feet could the plant grow in 0.5 day?

11. The distance from the park to the grocery store is 0.9 mile. Ezra runs 8 tenths of that distance and walks the rest of the way. How far does Ezra run from the park to the grocery store?

_____   _____

Lesson Check

1. Liz is hiking a trail that is 0.8 mile long. Liz hikes the first 2 tenths of the distance by herself. She hikes the rest of the way with her friends. How far does Liz hike by herself?

   (A) 0.16 mile

   (B) 0.20 mile

   (C) 0.80 mile

   (D) 1 mile

2. One cup of cooked zucchini has 1.9 grams of protein. How much protein is in 0.5 cup of zucchini?

   (A) 0.5 gram

   (B) 0.95 gram

   (C) 1.9 grams

   (D) 2.4 grams

Spiral Review

3. Which property does the statement show? (Lesson 1.3)

   $$(4 \times 8) \times 3 = (8 \times 4) \times 3$$

   (A) Commutative Property of Addition

   (B) Associative Property of Addition

   (C) Commutative Property of Multiplication

   (D) Associative Property of Multiplication

4. At the beginning of the school year, Rochelle joins the school garden club. In her plot of land, she plants 4 rows of tulips, each containing 27 bulbs. How many tulip bulbs does Rochelle plant in all? (Lesson 1.6)

   (A) 27   (C) 108

   (B) 88   (D) 216

5. In which place is the first digit of the quotient? (Lesson 2.1)

   $$3,589 \div 18$$

   (A) thousands

   (B) hundreds

   (C) tens

   (D) ones

6. At a football game, Jasmine bought a soft pretzel for $2.25 and a bottle of water for $1.50. She paid with a $5 bill. How much change should Jasmine get back? (Lesson 3.11)

   (A) $1.25

   (B) $1.50

   (C) $2.25

   (D) $3.75

Multiply Decimals

Find the product.

1.   5.8   58
    × 2.4   × 24
    13.92   **232**
              + 1,160
               1,392

2.   7.3
    × 9.6

3.   46.3
    × 0.8

4.   29.5
    × 1.3

5.   3.76
    × 4.8

6.   9.07
    × 6.5

7. 0.42 × 75.3

8. 5.6 × 61.84

9. 7.5 × 18.74

10. 0.9 × 53.8

Problem Solving  REAL WORLD

11. Aretha runs a marathon in 3.25 hours. Neal takes 1.6 times as long to run the same marathon. How many hours does it take Neal to run the marathon?

12. Tiffany catches a fish that weighs 12.3 pounds. Frank catches a fish that weighs 2.5 times as much as Tiffany's fish. How many pounds does Frank's fish weigh?

Lesson Check

1. Sue buys material to make a costume. She buys 1.75 yards of red material. She buys 1.2 times as many yards of blue material. How many yards of blue material does Sue buy?

   (A) 2.1 yards

   (B) 2.95 yards

   (C) 5.25 yards

   (D) 21 yards

2. Last week Juan worked 20.5 hours. This week he works 1.5 times as many hours as he did last week. How many hours does Juan work this week?

   (A) 12.3 hours

   (B) 22 hours

   (C) 30.75 hours

   (D) 37.5 hours

Spiral Review

3. The expression below shows a number in expanded form. What is the standard form of the number? (Lesson 3.2)

   $2 \times 10 + 3 \times \frac{1}{10} + 9 \times \frac{1}{100} + 7 \times \frac{1}{1,000}$

   (A) 2,397

   (B) 20.397

   (C) 2.397

   (D) 2.0397

4. Kelly buys a sweater for $16.79 and a pair of pants for $28.49. She pays with a $50 bill. How much change should Kelly get back? (Lesson 3.11)

   (A) $4.72

   (B) $5.48

   (C) $5.72

   (D) $45.28

5. Elvira is using a pattern to multiply $10^3 \times 37.2$.

   $10^0 \times 37.2 = 37.2$
   $10^1 \times 37.2 = 372$
   $10^2 \times 37.2 = \underline{\hspace{1cm}}$
   $10^3 \times 37.2 = \underline{\hspace{1cm}}$

   What is the product $10^3 \times 37.2$? (Lesson 4.1)

   (A) 0.0372

   (B) 0.372

   (C) 3,720

   (D) 37,200

6. Which digit should go in the box to make the following statement true? (Lesson 3.3)

   $63.749 < 63.\boxed{}2$

   (A) 3

   (B) 6

   (C) 7

   (D) 8

Name _____

Zeros in the Product

Find the product.

1.   0.07   7   2.   0.3   3.   0.05   4.   0.08
  × 0.2   × 2   × 0.1   × 0.8   × 0.3
  0.014   **14**

5.   0.06   6.   0.2   7.   0.05   8.   0.08
  × 0.7   × 0.4   × 0.4   × 0.8

9. $0.90   10.   0.02   11.   0.09   12. $0.05
  × 0.1   × 0.3   × 0.5   × 0.2

Problem Solving  REAL WORLD

13. A beaker contains 0.5 liter of a solution. Jordan uses 0.08 of the solution for an experiment. How much of the solution does Jordan use?

14. A certain type of nuts are on sale at $0.35 per pound. Tamara buys 0.2 pound of nuts. How much will the nuts cost?

_____   _____

Lesson Check

1. Cliff multiplies 0.06 and 0.5. Which product should he record?

   (A) 0.003

   (B) 0.03

   (C) 0.3

   (D) 3

2. Which product is equal to 0.036?

   (A) 0.6×0.6

   (B) 0.30×0.06

   (C) 0.9×0.4

   (D) 0.4×0.09

Spiral Review

3. A florist makes 24 bouquets. She uses 16 flowers for each bouquet. Altogether, how many flowers does she use? (Lesson 1.7)

   (A) 40

   (B) 168

   (C) 364

   (D) 384

4. Mark has 312 books in his bookcases. He has 11 times as many fiction books as nonfiction books. How many fiction books does Mark have? (Lesson 2.9)

   (A) 26

   (B) 212

   (C) 286

   (D) 301

5. Dwayne buys a pumpkin that weighs 12.65 pounds. To the nearest tenth of a pound, how much does the pumpkin weigh? (Lesson 3.4)

   (A) 12 pounds

   (B) 12.6 pounds

   (C) 12.7 pounds

   (D) 13 pounds

6. In which of the following numbers does the digit 6 have the value 6,000? (Lesson 1.2)

   (A) 896,000

   (B) 869,000

   (C) 809,600

   (D) 809,060

Name _____

Chapter 4 Extra Practice

Lesson 4.1

Complete the pattern.

1. $3.04 \times 1 =$ _____

$3.04 \times 10 =$ _____

$3.04 \times 100 =$ _____

$3.04 \times 1{,}000 =$ _____

2. $1 \times 70 =$ _____

$0.1 \times 70 =$ _____

$0.01 \times 70 =$ _____

3. $10^0 \times 0.57 =$ _____

$10^1 \times 0.57 =$ _____

$10^2 \times 0.57 =$ _____

$10^3 \times 0.57 =$ _____

Lesson 4.2

Use the decimal model to find the product.

1. $4 \times 0.07 =$ _____

2. $5 \times 0.20 =$ _____

3. $13 \times 0.03 =$ _____

Find the product. Draw a quick picture.

4. $7 \times 0.06 =$ _____

5. $8 \times 0.12 =$ _____

Lessons 4.3 - 4.4, 4.7 - 4.8

Find the product.

1. 8.24
    ×   8

2. 5.3
    × 7

3. 0.29
    ×   4

4.   18.5
    × 0.42

5.   4.4
    × 8.7

6.   9.2
    × 2.8

7. 0.04
    × 0.3

8. $0.70
    × 0.6

9. 17 × 0.16

10. 3.55 × 75.2

11. 7.8 × 25.87

Lesson 4.5

Solve.

1. Some friends go to the store to buy school supplies. Noel spends $4.89. Holly spends 3 times as much as Noel. Kris spends $12.73 more than Holly. How much does Kris spend?

2. Chase collects $27.34 for a fundraiser. Sydney collects $9.83 more than Chase. Ally collects 4 times as much as Sydney. How much money does Ally collect?

Lesson 4.6

Multiply. Use the decimal model.

1. 0.4 × 0.7 = _____

2. 0.9 × 0.8 = _____

3. 0.3 × 1.6 = _____

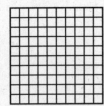

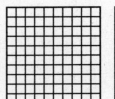
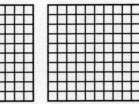

School-Home Letter

© Houghton Mifflin Harcourt Publishing Company

Vocabulary

decimal A number with one or more digits to the right of the decimal point

dividend The number that is to be divided in a division problem

divisor The number that divides the dividend

quotient The number that results from dividing

Dear Family,

Throughout the next few weeks, our math class will be learning about decimal division. We will also be learning how to estimate decimal quotients.

You can expect to see homework that involves division of decimals through hundredths.

Here is a sample of how your child is taught to divide decimals.

🔑 MODEL Divide Decimals

Divide. 44.8 ÷ 3.2

STEP 1

Estimate.

$45 \div 3 = 15$

STEP 2

Make the divisor a whole number by multiplying the divisor and dividend by the same power of 10.

$$3.2\overline{)44.8}$$

STEP 3

Divide.

$$\begin{array}{r} 14 \\ 32\overline{)448} \\ -32 \\ \hline 128 \\ -128 \\ \hline 0 \end{array}$$

So, 44.8 ÷ 3.2 = 14.

Tips

Estimating with Decimals

When estimating, it may be helpful to round the numbers in the problem to compatible numbers. Compatible numbers are pairs of numbers that are easy to compute with mentally.

For example, to estimate 19.68 ÷ 4.1, use the compatible numbers 20 and 4: 20 ÷ 4 = 5.

Activity

Use trips to grocery or department stores as opportunities to practice decimal division. For example, "Which is the better buy, the 10-ounce box of cereal for $3.25 or the 15-ounce box for $4.65?" Work together to write a division sentence to represent each situation. Help your child estimate the quotient and then find the exact answer.

5

Carta para la casa

Querida familia,

Durante las próximas semanas, en la clase de matemáticas aprenderemos la división decimal. También aprenderemos a estimar cocientes decimales.

Llevaré a casa tareas con actividades que incluyan división de decimales hasta las centésimas.

Este es un ejemplo de cómo vamos a dividir decimales.

Vocabulario

decimal Un número con uno o más dígitos a la derecha del punto decimal

dividendo El número que se va a dividir en un problema de división

divisor El número que divide al dividendo

cociente El número que se obtiene al resolver una división

🔑 MODELO Dividir decimales

Divide $44.8 \div 3.2$

PASO 1

Estima. $45 \div 3 = 15$

PASO 2

Convierte el divisor a un número entero multiplicando el divisor y el dividendo por la misma potencia de 10.

$3.2\overline{)44.8}$

PASO 3

Divide.

$$\begin{array}{r} 14 \\ 32\overline{)448} \\ -32 \\ \hline 128 \\ -128 \\ \hline 0 \end{array}$$

Por lo tanto, $44.8 \div 3.2 = 14$.

Pistas

Estimar con decimales

Al estimar, puede resultar útil redondear los números del problema a números compatibles. Los números compatibles son pares de números que son fáciles de computar mentalmente.

Por ejemplo, para estimar $19.68 \div 4.1$, usa los números compatibles 20 y 4: $20 \div 4 = 5$.

Actividad

Use los paseos a las tiendas de alimentos o departamentales para practicar la división decimal. Por ejemplo, "¿Qué conviene comprar, la caja de cereales de 10 onzas por $3.25 o la caja de 15 onzas por $4.65?" Trabajen juntos para escribir el enunciado de división que represente cada situación. Ayude a su hijo o hija a estimar el cociente y luego hallen la respuesta exacta.

Name _____

Division Patterns with Decimals

Complete the pattern.

1. $78.3 \div 1 =$ ___**78.3**___

   $78.3 \div 10 =$ ___**7.83**___

   $78.3 \div 100 =$ **0.783**

2. $179 \div 10^0 =$ _____

   $179 \div 10^1 =$ _____

   $179 \div 10^2 =$ _____

   $179 \div 10^3 =$ _____

3. $87.5 \div 10^0 =$ _____

   $87.5 \div 10^1 =$ _____

   $87.5 \div 10^2 =$ _____

4. $124 \div 1 =$ _____

   $124 \div 10 =$ _____

   $124 \div 100 =$ _____

   $124 \div 1,000 =$ _____

5. $18 \div 1 =$ _____

   $18 \div 10 =$ _____

   $18 \div 100 =$ _____

   $18 \div 1,000 =$ _____

6. $23 \div 10^0 =$ _____

   $23 \div 10^1 =$ _____

   $23 \div 10^2 =$ _____

   $23 \div 10^3 =$ _____

7. $51.8 \div 1 =$ _____

   $51.8 \div 10 =$ _____

   $51.8 \div 100 =$ _____

8. $49.3 \div 10^0 =$ _____

   $49.3 \div 10^1 =$ _____

   $49.3 \div 10^2 =$ _____

9. $32.4 \div 10^0 =$ _____

   $32.4 \div 10^1 =$ _____

   $32.4 \div 10^2 =$ _____

Problem Solving  REAL WORLD

10. The local café uses 510 cups of mixed vegetables to make 1,000 quarts of beef barley soup. Each quart of soup contains the same amount of vegetables. How many cups of vegetables are in each quart of soup?

    _____

11. The same café uses 18.5 cups of flour to make 100 servings of pancakes. How many cups of flour are in one serving of pancakes?

    _____

Lesson Check

1. The Statue of Liberty is 305.5 feet tall, from the foundation of its pedestal to the top of its torch. Isla is building a model of the statue. The model will be one-hundredth as tall as the actual statue. How tall will the model be?

   (A)   0.3055 foot

   (B)   3.055 feet

   (C)   30.55 feet

   (D) 30,550 feet

2. Sue's teacher asked her to find $42.6 \div 10^2$. How should Sue move the decimal point to get the correct quotient?

   (A) 2 places to the right

   (B) 1 place to the right

   (C) 1 place to the left

   (D) 2 places to the left

Spiral Review

3. In the number 956,783,529, how does the value of the digit 5 in the ten millions place compare to the digit 5 in the hundreds place? (Lesson 1.2)

   (A)   100 times as much as

   (B)   1,000 times as much as

   (C)   10,000 times as much as

   (D) 100,000 times as much as

4. Taylor has $97.23 in her checking account. She uses her debit card to spend $29.74 and then deposits $118.08 into her account. What is Taylor's new balance? (Lesson 3.11)

   (A) $8.89

   (B) $185.57

   (C) $215.31

   (D) $245.05

5. At the bank, Brent exchanges $50 in bills for 50 one-dollar coins. The total mass of the coins is 405 grams. Estimate the mass of 1 one-dollar coin. (Lesson 2.5)

   (A)   1 gram

   (B)   8 grams

   (C)   50 grams

   (D) 100 grams

6. A commercial jetliner has 245 passenger seats. The seats are arranged in 49 equal rows. How many seats are in each row? (Lesson 2.6)

   (A)   5

   (B)   49

   (C)   245

   (D) 1,225

Name _____

Divide Decimals by Whole Numbers

Use the model to complete the number sentence.

1. 1.2 ÷ 4 = __0.3__

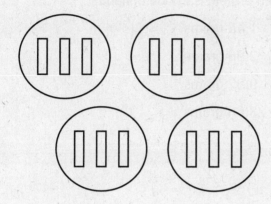

2. 3.69 ÷ 3 = _____

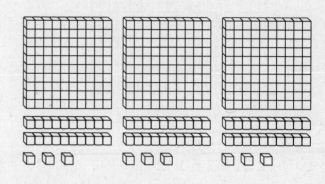

Divide. Use base-ten blocks.

3. 4.9 ÷ 7 = _____

4. 3.6 ÷ 9 = _____

5. 2.4 ÷ 8 = _____

6. 6.48 ÷ 4 = _____

7. 3.01 ÷ 7 = _____

8. 4.26 ÷ 3 = _____

Problem Solving REAL WORLD

9. In PE class, Carl runs a distance of 1.17 miles in 9 minutes. At that rate, how far does Carl run in one minute?

10. Marianne spends $9.45 on 5 greeting cards. Each card costs the same amount. What is the cost of one greeting card?

_____

_____

Lesson Check

1. Which division sentence does the model below represent?

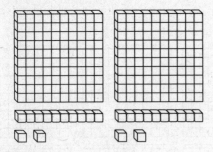

   (A) $1.12 \div 2 = 0.56$

   (B) $2.24 \div 2 = 1.12$

   (C) $2.24 \div 4 = 0.56$

   (D) $3.36 \div 3 = 1.12$

2. A bunch of 4 bananas contains a total of 5.92 grams of protein. Suppose each banana contains the same amount of protein. How much protein is in one banana?

   (A) 1.48 grams

   (B) 2.96 grams

   (C) 9.92 grams

   (D) 23.68 grams

Spiral Review

3. At the deli, one pound of turkey costs $7.98. Mr. Epstein buys 3 pounds of turkey. How much will the turkey cost? **(Lesson 4.5)**

   (A) $2.66

   (B) $7.98

   (C) $15.96

   (D) $23.94

4. Mrs. Cho drives 45 miles in 1 hour. If her speed stays constant, how many hours will it take for her to drive 405 miles? **(Lesson 2.6)**

   (A) 360 hours

   (B) 45 hours

   (C) 9 hours

   (D) 8 hours

5. Which list shows the following numbers in order from least to greatest? **(Lesson 3.3)**

   1.23; 1.2; 2.31; 3.2

   (A) 1.23, 1.2, 2.31, 3.2

   (B) 1.2, 1.23, 2.31, 3.2

   (C) 3.2, 2.31, 1.23, 1.2

   (D) 1.2, 1.23, 3.2, 2.31

6. Over the weekend, Aiden spent 15 minutes on his math homework. He spent three times as much time on his science homework. How much time did Aiden spend on his science homework? **(Lesson 1.6)**

   (A) 5 minutes

   (B) 15 minutes

   (C) 30 minutes

   (D) 45 minutes

Estimate Quotients

Use compatible numbers to estimate the quotient.

1. $19.7 \div 3$

$$18 \div 3 = 6$$

2. $394.6 \div 9$

3. $308.3 \div 15$

Estimate the quotient.

4. $63.5 \div 5$

5. $57.8 \div 81$

6. $172.6 \div 39$

7. $43.6 \div 8$

8. $2.8 \div 6$

9. $467.6 \div 8$

10. $209.3 \div 48$

11. $737.5 \div 9$

12. $256.1 \div 82$

Problem Solving

13. Taylor uses 645.6 gallons of water in 7 days. Suppose he uses the same amount of water each day. About how much water does Taylor use each day?

14. On a road trip, Sandy drives 368.7 miles. Her car uses a total of 18 gallons of gas. About how many miles per gallon does Sandy's car get?

_____

_____

Lesson Check

1. Terry bicycled 64.8 miles in 7 hours. Which is the best estimate of the average number of miles she bicycled each hour?

   (A) about 0.8 mile

   (B) about 0.9 mile

   (C) about 8 miles

   (D) about 9 miles

2. Which is the best estimate for the following quotient?

   $$891.3 \div 28$$

   (A) about 3

   (B) about 4

   (C) about 30

   (D) about 40

Spiral Review

3. An object that weighs 1 pound on Earth weighs 1.19 pounds on Neptune. Suppose a dog weighs 9 pounds on Earth. How much would the same dog weigh on Neptune? (Lesson 4.3)

   (A)   0.1071 pound

   (B)   1.071 pounds

   (C) 10.71 pounds

   (D) 107.1 pounds

4. A bookstore orders 200 books. The books are packaged in boxes that hold 24 books each. All the boxes the bookstore receives are full, except one. How many boxes does the bookstore receive? (Lesson 2.7)

   (A) 8

   (B) 9

   (C) 10

   (D) 11

5. Tara has $2,000 in her savings account. David has one-tenth as much as Tara in his savings account. How much does David have in his savings account? (Lesson 4.1)

   (A) $2

   (B) $20

   (C) $200

   (D) $20,000

6. Which of the following statements is true? (Lesson 3.3)

   (A) 7.63 > 7.629

   (B) 5.134 > 5.14

   (C) 8.23 < 8.230

   (D) 4.079 = 4.790

Division of Decimals by Whole Numbers

Divide.

1.
```
     1.32
  7)9.24
   -7
   ──
   22
  -21
   ──
    14
   -14
   ───
     0
```

2. 6)5.04

3. 23)85.1

4. 36)86.4

5. 6)$6.48

6. 8)59.2

7. 5)2.35

8. 41)278.8

9. 19)$70.49

10. 4)$9.48

11. 18)82.8

12. 37)32.93

Problem Solving  REAL WORLD

13. On Saturday, 12 friends go ice skating. Altogether, they pay $83.40 for admission. They share the cost equally. How much does each person pay?

14. A team of 4 people participates in a 400-yard relay race. Each team member runs the same distance. The team completes the race in a total of 53.2 seconds. What is the average running time for each person?

_____

_____

Lesson Check

1. Theresa pays $9.56 for 4 pounds of tomatoes. What is the cost of 1 pound of tomatoes?

Ⓐ $0.24

Ⓑ $2.39

Ⓒ $23.90

Ⓓ $38.24

2. Robert wrote the division problem below. What is the quotient?

$$13\overline{)83.2}$$

Ⓐ 6.4

Ⓑ 6.6

Ⓒ 64

Ⓓ 66

Spiral Review

3. What is the value of the following expression? (Lesson 1.12)

$$2 \times \{6 + [12 \div (3 + 1)]\} - 1$$

Ⓐ 13

Ⓑ 17

Ⓒ 18

Ⓓ 21

4. Last month, Dory biked 11 times as many miles as Karly. Together they biked a total of 156 miles. How many miles did Dory bike last month? (Lesson 2.9)

Ⓐ 11 miles

Ⓑ 13 miles

Ⓒ 142 miles

Ⓓ 143 miles

5. Jin ran 15.2 miles over the weekend. He ran 6.75 miles on Saturday. How many miles did he run on Sunday? (Lesson 3.9)

Ⓐ 8.45 miles

Ⓑ 8.55 miles

Ⓒ 9.45 miles

Ⓓ 9.55 miles

6. A bakery used 475 pounds of apples to make 1,000 apple tarts. Each tart contains the same amount of apples. How many pounds of apples are used in each tart? (Lesson 5.1)

Ⓐ 47.5 pounds

Ⓑ 4.75 pounds

Ⓒ 0.475 pound

Ⓓ 0.0475 pound

Name _____

Decimal Division

Use the model to complete the number sentence.

1. $1.6 \div 0.4 =$ _____**4**_____

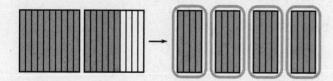

2. $0.36 \div 0.06 =$ _____

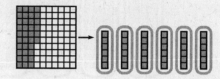

Divide. Use decimal models.

3. $2.8 \div 0.7 =$ _____

4. $0.40 \div 0.05 =$ _____

5. $0.45 \div 0.05 =$ _____

6. $1.62 \div 0.27 =$ _____

7. $0.56 \div 0.08 =$ _____

8. $1.8 \div 0.9 =$ _____

Problem Solving

9. Keisha buys 2.4 kilograms of rice. She separates the rice into packages that contain 0.4 kilogram of rice each. How many packages of rice can Keisha make?

_____

10. Leighton is making cloth headbands. She has 4.2 yards of cloth. She uses 0.2 yard of cloth for each headband. How many headbands can Leighton make from the length of cloth she has?

_____

Lesson Check

1. What number sentence does the model represent?

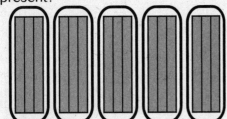

- (A) $1.5 \div 0.3 = 5$
- (B) $1.5 \div 0.5 = 3$
- (C) $0.9 \div 0.3 = 3$
- (D) $2.5 \div 0.5 = 5$

2. Morris has 1.25 pounds of strawberries. He uses 0.25 pound of strawberries to make one serving. How many servings can Morris make?

- (A) 0.25
- (B) 0.5
- (C) 1.25
- (D) 5

Spiral Review

3. Which property does the following equation show? (Lesson 1.3)

$$5 + 7 + 9 = 7 + 5 + 9$$

- (A) Commutative Property of Addition
- (B) Associative Property of Addition
- (C) Commutative Property of Multiplication
- (D) Associative Property of Multiplication

4. An auditorium has 25 rows with 45 seats in each row. How many seats are there in all? (Lesson 1.7)

- (A) 25
- (B) 45
- (C) 125
- (D) 1,125

5. Volunteers at an animal shelter divided 132 pounds of dry dog food equally into 16 bags. How many pounds of dog food did they put in each bag? (Lesson 2.7)

- (A) 8 pounds
- (B) $8\frac{1}{4}$ pounds
- (C) $8\frac{1}{2}$ pounds
- (D) 9 pounds

6. At the movies, Aaron buys popcorn for $5.25 and a bottle of water for $2.50. He pays with a $10 bill. How much change should Aaron receive? (Lesson 3.11)

- (A) $2.25
- (B) $2.50
- (C) $5.25
- (D) $7.75

Name _____

Divide Decimals

Divide.

1. $0.4\overline{)8.4}$

Multiply both 0.4 and 8.4 by 10 to make the divisor a whole number. Then divide.

$$\begin{array}{r} 21 \\ 4\overline{)84} \\ -8 \\ \hline 04 \\ -4 \\ \hline 0 \end{array}$$

2. $0.2\overline{)0.4}$

3. $0.07\overline{)1.68}$

4. $0.37\overline{)5.18}$

5. $0.4\overline{)10.4}$

6. $6.3 \div 0.7$

7. $1.52 \div 1.9$

8. $12.24 \div 0.34$

9. $10.81 \div 2.3$

Problem Solving  REAL WORLD

10. At the market, grapes cost $0.85 per pound. Clarissa buys grapes and pays a total of $2.55. How many pounds of grapes does she buy?

11. Damon kayaks on a river near his home. He plans to kayak a total of 6.4 miles. Damon kayaks at an average speed of 1.6 miles per hour. How many hours will it take Damon to kayak the 6.4 miles?

_____

_____

Lesson Check

1. Lee walked a total of 4.48 miles. If he walks 1.4 miles each hour. How long did Lee walk?

   (A) 3.08 hours

   (B) 3.2 hours

   (C) 6.272 hours

   (D) 32 hours

2. Janelle has 3.6 yards of wire, which she wants to use to make bracelets. She needs 0.3 yard for each bracelet. Altogether, how many bracelets can Janelle make?

   (A) 1.08

   (B) 3.3

   (C) 3.9

   (D) 12

Spiral Review

3. Susie's teacher asks her to complete the multiplication problem below. What is the product? (Lesson 4.7)

$$\begin{array}{r} 0.3 \\ \times\ \ 3.7 \\ \hline \end{array}$$

   (A) 0.111

   (B) 1.11

   (C) 11.1

   (D) 111

4. At an Internet store, a laptop computer costs $724.99. At a local store, the same computer costs $879.95. What is the difference in prices? (Lesson 3.9)

   (A) $154.96

   (B) $155.04

   (C) $155.16

   (D) $155.96

5. Continue the pattern below. What is the quotient $75.8 \div 10^2$? (Lesson 5.1)

   $75.8 \div 10^0 = 75.8$

   $75.8 \div 10^1 = \underline{\hphantom{aaaa}}$

   $75.8 \div 10^2 = \underline{\hphantom{aaaa}}$

   (A) 0.758

   (B) 7.58

   (C) 758

   (D) 7,580

6. Which number will make the following statement true? (Lesson 3.3)

   $58.827 < 58.\square 1$

   (A) 2

   (B) 3

   (C) 8

   (D) 9

Name _____

Write Zeros in the Dividend

Divide.

1.
```
      3.95
  6)23.70
  −18
  ───
    57
   −54
   ───
     30
    −30
    ───
      0
```

2. 25)405

3. 0.6)12.9

4. 0.8)30

5. 4)36.2

6. 35)97.3

7. 7.8 ÷ 15

8. 49 ÷ 14

9. 52.2 ÷ 12

10. 1.14 ÷ 0.76

11. 20.2 ÷ 4

12. 138.4 ÷ 16

Problem Solving REAL WORLD

13. Mark has a board that is 12 feet long. He cuts the board into 8 pieces that are the same length. How long is each piece?

14. Josh pays $7.59 for 2.2 pounds of ground turkey. What is the price per pound of the ground turkey?

Lesson Check

1. Tina divides 21.4 ounces of trail mix equally into 5 bags. How many ounces of trail mix are in each bag?

   (A) 0.428 ounce

   (B) 4.28 ounces

   (C) 42.8 ounces

   (D) 428 ounces

2. A slug crawls 5.62 meters in 0.4 hours. What is the slug's speed in meters per hour?

   (A) 0.1405 meters per hour

   (B) 1.405 meters per hour

   (C) 14.05 meters per hour

   (D) 140.5 meters per hour

Spiral Review

3. Suzy buys 35 pounds of rice. She divides it equally into 100 bags. How many pounds of rice does Suzy put in each bag? (Lesson 5.1)

   (A) 0.035 pound

   (B) 0.35 pound

   (C) 3.5 pounds

   (D) 3,500 pounds

4. Juliette spends $6.12 at the store. Morgan spends 3 times as much as Juliette. Jonah spends $4.29 more than Morgan. How much money does Jonah spend? (Lesson 4.5)

   (A) $31.23

   (B) $22.65

   (C) $10.41

   (D) $5.49

5. A concert sold out for 12 performances. Altogether, 8,208 tickets were sold. How many tickets were sold for each performance? (Lesson 2.6)

   (A) 679

   (B) 684

   (C) 689

   (D) 694

6. Jared has two dogs, Spot and Rover. Spot weighs 75.25 pounds. Rover weighs 48.8 pounds more than Spot. How much does Rover weigh? (Lesson 3.8)

   (A) 34.45 pounds

   (B) 123.33 pounds

   (C) 124.05 pounds

   (D) 124.5 pounds

Name _____

Problem Solving • Decimal Operations

1. Lily spent $30.00 on a T-shirt, a sandwich, and 2 books. The T-shirt cost $8.95, and the sandwich cost $7.25. The books each cost the same amount. How much did each book cost?

(2 × cost of each book) + $8.95 + $7.25 = $30.00

$30.00 − $8.95 − $7.25 = (2 × cost of each book)

(2 × cost of each book) = $13.80
$13.80 ÷ 2 = $6.90

$6.90

2. Meryl spends a total of $68.82 for 2 pairs of sneakers with the same cost. The sales tax is $5.32. Meryl also uses a coupon for $3.00 off her purchase. How much does each pair of sneakers cost?

_____

3. A 6-pack of undershirts costs $13.98. This is $3.96 less than the cost of buying 6 individual shirts. If each undershirt costs the same amount, how much does each undershirt cost when purchased individually?

_____

4. Mason spent $15.85 for 3 notebooks and 2 boxes of markers. The boxes of markers cost $3.95 each, and the sales tax was $1.23. Mason also used a coupon for $0.75 off his purchase. If each notebook had the same cost, how much did each notebook cost?

_____

Lesson Check

1. Joe spends $8 on lunch and $6.50 on dry cleaning. He also buys 2 shirts that each cost the same amount. Joe spends a total of $52. What is the cost of each shirt?

   (A) $18.25

   (B) $18.75

   (C) $33.75

   (D) $37.50

2. Tina uses a $50 gift certificate to buy a pair of pajamas for $17.97, a necklace for $25.49, and 3 pairs of socks that each cost the same amount. Tina has to pay $0.33 because the gift certificate does not cover the total cost of all the items. How much does each pair of socks cost?

   (A) $0.11

   (B) $2.07

   (C) $2.18

   (D) $2.29

Spiral Review

3. Which list orders the numbers from least to greatest? (Lesson 3.3)

   (A) 0.123, 2.13, 3.12, 2.31

   (B) 0.123, 2.13, 2.31, 3.12

   (C) 2.13, 0.123, 3.12, 2.31

   (D) 3.12, 2.31, 2.13, 0.123

4. Stephen wrote the problem $46.8 \div 0.5$. What is the correct quotient? (Lesson 5.7)

   (A) 0.936

   (B) 9.36

   (C) 93.6

   (D) 936

5. Sarah, Juan, and Larry are on the track team. Last week, Sarah ran 8.25 miles, Juan ran 11.8 miles, and Larry ran 9.3 miles. How many miles did they run altogether? (Lesson 3.8)

   (A) 28.35 miles

   (B) 28.36 miles

   (C) 29.35 miles

   (D) 29.36 miles

6. On a fishing trip, Lucy and Ed caught one fish each. Ed's fish weighed 6.45 pounds. Lucy's fish weighed 1.6 times as much. How much did Lucy's fish weigh? (Lesson 4.7)

   (A) 4.85 pounds

   (B) 8.05 pounds

   (C) 10.32 pounds

   (D) 103.20 pounds

Name _____

Chapter 5 Extra Practice

Lesson 5.1

Complete the pattern.

1. $274 \div 1 =$ _____

$274 \div 10 =$ _____

$274 \div 100 =$ _____

$274 \div 1,000 =$ _____

2. $83 \div 1 =$ _____

$83 \div 10 =$ _____

$83 \div 100 =$ _____

$83 \div 1,000 =$ _____

3. $12 \div 10^0 =$ _____

$12 \div 10^1 =$ _____

$12 \div 10^2 =$ _____

$12 \div 10^3 =$ _____

Lessons 5.2, 5.4

Use the model to complete the number sentence.

1. $1.2 \div 4 =$ _____

2. $3.75 \div 3 =$ _____

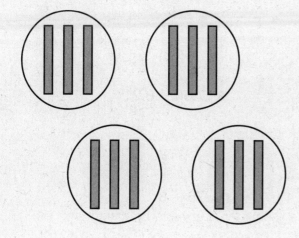

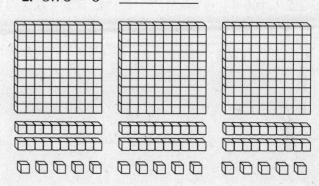

Divide.

3. $2.4 \div 6 =$ _____

4. $4.9 \div 7 =$ _____

5. $4.92 \div 4 =$ _____

6. $7\overline{)9.24}$

7. $4\overline{)\$7.64}$

8. $52\overline{)140.4}$

Lesson 5.3

Estimate the quotient.

1. $28.3 \div 9$ **2.** $74.3 \div 8$ **3.** $198.4 \div 21$

Lessons 5.5 - 5.7

Use the model to complete the number sentence.

1. $1.60 \div 0.8 =$ _____   **2.** $0.32 \div 0.08 =$ _____

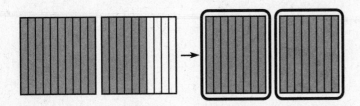 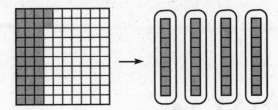

Divide.

3. $3.2 \div 0.8$ **4.** $8.1 \div 0.9$ **5.** $1.68 \div 0.56$

6. $0.7 \overline{)16.1}$ **7.** $5.6 \div 0.7$ **8.** $7.56 \div 0.21$

9. $8 \overline{)92}$ **10.** $11 \div 2.5$ **11.** $76 \div 8$

Lesson 5.8

1. Quinn spent $16.49 for 3 magazines and 4 sheets of stickers. The magazines cost $3.99 each and the sales tax was $1.02. Quinn also used a coupon for $1.50 off her purchase. If each sheet of stickers had the same cost, how much did each sheet of stickers cost?

2. Jairo spent $40.18 on 3 music CDs. Each CD cost the same amount. The sales tax was $2.33. Jairo also used a coupon for $1.00 off his purchase. How much did each CD cost?

School-Home Letter

© Houghton Mifflin Harcourt Publishing Company

Vocabulary

common denominator A common multiple of two or more denominators

equivalent fractions Fractions that name the same amount or part

least common multiple The least number that is a common multiple of two or more numbers

least common denominator The least common multiple of two or more denominators

Dear Family,

Throughout the next few weeks, our math class will study the operations of addition and subtraction with fractions. The students will study and learn to identify and apply common denominators.

You can expect to see homework that includes adding and subtracting mixed numbers.

Here is a sample of how your child will be taught to estimate differences of fractions.

🔑 MODEL Estimate Sums and Differences

This is how we will be estimating $\frac{7}{8} - \frac{2}{5}$.

STEP 1

Round each fraction to the nearest 0, $\frac{1}{2}$, or 1.

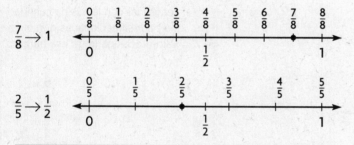

Tips

Subtraction with Renaming

If you have to regroup from the whole number in a mixed number, remember to regroup into the same fractional parts as in that number's fraction value.

STEP 2

Subtract the rounded fractions.

$1 - \frac{1}{2} = \frac{1}{2}$

Activity

Having a ruler handy helps to quickly identify benchmarks to use when estimating fraction sums and differences. Have your child estimate fraction sums and differences such as $\frac{7}{8} - \frac{2}{4}$ and $1\frac{3}{8} + \frac{1}{4}$.

Carta para la casa

© Houghton Mifflin Harcourt Publishing Company

Vocabulario

común denominador Un múltiplo común de dos o más denominadores

fracciones equivalentes Fracciones que nombran la misma cantidad o parte

mínimo común múltiplo El menor número, que es un múltiplo común de dos o más números

mínimo común denominador El menor múltiplo común de dos o más denominadores

Querida familia,

Durante las próximas semanas, en la clase de matemáticas estudiaremos las operaciones de suma y resta con fracciones. Aprenderemos a identificar y usar denominadores comunes.

Llevaré a la casa tareas con actividades para sumar y restar números mixtos.

Este es un ejemplo de la manera como aprenderemos a estimar diferencias de fracciones.

🔑 MODELO Estimar sumas y diferencias

Así es como estimaremos $\frac{7}{8} - \frac{2}{5}$.

Pistas

Restar con conversión

Para reagrupar el número entero como un número mixto, recuerda reagrupar con la misma cantidad de partes fraccionarias que el valor fraccionario de ese número.

PASO 1

Redondea cada fracción al 0, al $\frac{1}{2}$ o al 1 más cercano.

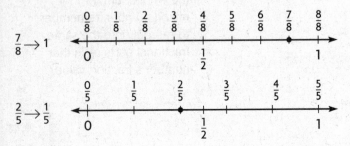

PASO 2

Resta las fracciones redondeadas.

$$1 - \frac{1}{2} = \frac{1}{2}$$

Actividad

Una regla puede servir para identificar rápidamente los puntos de referencia que se usan cuando se estiman sumas y diferencias de fracciones. Pida a su hijo que estime sumas y diferencias de fracciones como $\frac{7}{8} - \frac{2}{4}$ y $1\frac{3}{8} + \frac{1}{4}$.

Name _____

Addition with Unlike Denominators

Use fraction strips to find the sum. Write your answer in simplest form.

1. $\frac{1}{2} + \frac{3}{4}$

   $\frac{1}{2} + \frac{3}{4} = \frac{2}{4} + \frac{3}{4} = \frac{5}{4}$, or $1\frac{1}{4}$

   $1\frac{1}{4}$
   _____

2. $\frac{1}{3} + \frac{1}{4}$

   _____

3. $\frac{3}{5} + \frac{1}{2}$

   _____

4. $\frac{3}{8} + \frac{1}{2}$

   _____

5. $\frac{1}{4} + \frac{5}{8}$

   _____

6. $\frac{2}{3} + \frac{3}{4}$

   _____

7. $\frac{1}{2} + \frac{2}{5}$

   _____

8. $\frac{2}{3} + \frac{1}{2}$

   _____

9. $\frac{7}{8} + \frac{1}{2}$

   _____

10. $\frac{5}{6} + \frac{1}{3}$

    _____

11. $\frac{1}{5} + \frac{1}{2}$

    _____

12. $\frac{3}{4} + \frac{3}{8}$

    _____

Problem Solving  REAL WORLD

13. Brandus bought $\frac{1}{3}$ pound of ground turkey and $\frac{3}{4}$ pound of ground beef to make sausages. How many pounds of meat did he buy?

    _____

14. To make a ribbon and bow for a hat, Stacey needs $\frac{5}{6}$ yard of black ribbon and $\frac{2}{3}$ yard of red ribbon. How much total ribbon does she need?

    _____

Lesson Check

1. Hirva ate $\frac{5}{8}$ of a medium pizza. Elizabeth ate $\frac{1}{4}$ of the pizza. How much pizza did they eat altogether?

   (A) $\frac{2}{4}$

   (B) $\frac{6}{12}$

   (C) $\frac{6}{8}$

   (D) $\frac{7}{8}$

2. Bill ate $\frac{1}{4}$ pound of trail mix on his first break during a hiking trip. On his second break, he ate $\frac{1}{6}$ pound. How many pounds of trail mix did he eat during both breaks?

   (A) $\frac{5}{6}$ pound

   (B) $\frac{5}{12}$ pound

   (C) $\frac{1}{3}$ pound

   (D) $\frac{1}{5}$ pound

Spiral Review

3. In 782,341,693, which digit is in the ten thousands place? (Lesson 1.1)

   (A) 2

   (B) 4

   (C) 8

   (D) 9

4. Matt ran 8 laps in 1,256 seconds. If he ran each lap in the same amount of time, how many seconds did it take him to run 1 lap? (Lesson 1.9)

   (A) 107 seconds

   (B) 132 seconds

   (C) 157 seconds

   (D) 170 seconds

5. Gilbert bought 3 shirts for $15.90 each, including tax. How much did he spend? (Lesson 4.3)

   (A) $5.30

   (B) $35.70

   (C) $37.70

   (D) $47.70

6. Julia has 14 pounds of nuts. There are 16 ounces in one pound. How many ounces of nuts does she have? (Lesson 1.7)

   (A) 224 ounces

   (B) 124 ounces

   (C) 98 ounces

   (D) 30 ounces

Name _____

Subtraction with Unlike Denominators

Use fraction strips to find the difference. Write your answer in simplest form.

1. $\frac{1}{2} - \frac{1}{3}$

$\frac{1}{2} - \frac{1}{3} = \frac{3}{6} - \frac{2}{6} = \frac{1}{6}$

$\frac{1}{6}$ _____

2. $\frac{3}{4} - \frac{3}{8}$

_____

3. $\frac{7}{8} - \frac{1}{2}$

_____

4. $\frac{1}{2} - \frac{1}{5}$

_____

5. $\frac{2}{3} - \frac{1}{4}$

_____

6. $\frac{4}{5} - \frac{1}{2}$

_____

7. $\frac{3}{4} - \frac{1}{3}$

_____

8. $\frac{5}{8} - \frac{1}{2}$

_____

9. $\frac{7}{10} - \frac{1}{2}$

_____

10. $\frac{9}{10} - \frac{2}{5}$

_____

11. $\frac{5}{8} - \frac{1}{4}$

_____

12. $\frac{2}{3} - \frac{1}{2}$

_____

Problem Solving REAL WORLD

13. Amber had $\frac{3}{8}$ of a cake left after her party. She wrapped a piece that was $\frac{1}{4}$ of the original cake for her best friend. What fractional part did she have left for herself?

_____

14. Wesley bought $\frac{1}{2}$ pound of nails for a project. When he finished the project, he had $\frac{1}{4}$ pound of the nails left. How many pounds of nails did he use?

_____

Lesson Check

1. A meatloaf recipe calls for $\frac{7}{8}$ cup of bread crumbs for the loaf and the topping. If $\frac{3}{4}$ cup is used for the loaf, what fraction of a cup is used for the topping?

   (A) $\frac{4}{4}$ cup

   (B) $\frac{4}{8}$ cup

   (C) $\frac{1}{4}$ cup

   (D) $\frac{1}{8}$ cup

2. Hannah bought $\frac{3}{4}$ yard of felt for a project. She used $\frac{1}{8}$ yard. What fraction of a yard of felt did she have left over?

   (A) $\frac{2}{8}$ yard

   (B) $\frac{4}{8}$ yard

   (C) $\frac{5}{8}$ yard

   (D) $\frac{5}{4}$ yards

Spiral Review

3. Jasmine's race time was 34.287 minutes. Round her race time to the nearest tenth of a minute. (Lesson 3.4)

   (A) 34.3 minutes

   (B) 34.2 minutes

   (C) 34.0 minutes

   (D) 30.0 minutes

4. The Art Club is having a fund-raiser, and 198 people are attending. If 12 people can sit at each table, what is the least number of tables needed? (Lesson 2.7)

   (A) 15

   (B) 16

   (C) 17

   (D) 20

5. During the day, Sam spent $4.85 on lunch. He also bought 2 books for $7.95 each. At the end of the day, he had $8.20 left. How much money did he start with? (Lesson 4.5)

   (A) $12.80

   (B) $20.75

   (C) $21.00

   (D) $28.95

6. What is the product of 7.5 and 1,000? (Lesson 4.1)

   (A) 0.0075

   (B) 0.075

   (C) 7,500

   (D) 75,000

Name _____

Estimate Fraction Sums and Differences

Estimate the sum or difference.

1. $\frac{1}{2} - \frac{1}{3}$

Think: $\frac{1}{3}$ is closer to $\frac{1}{2}$ than to 0.

Estimate: ___**0**___

2. $\frac{1}{8} + \frac{1}{4}$

Estimate: _____

3. $\frac{4}{5} - \frac{1}{2}$

Estimate: _____

4. $2\frac{3}{5} - 1\frac{3}{8}$

Estimate: _____

5. $\frac{1}{5} + \frac{3}{7}$

Estimate: _____

6. $\frac{2}{5} + \frac{2}{3}$

Estimate: _____

7. $2\frac{2}{3} + \frac{3}{4}$

Estimate: _____

8. $1\frac{7}{8} - 1\frac{1}{2}$

Estimate: _____

9. $4\frac{1}{8} - \frac{3}{4}$

Estimate: _____

10. $3\frac{9}{10} - 1\frac{2}{5}$

Estimate: _____

11. $2\frac{5}{8} + 1\frac{1}{4}$

Estimate: _____

12. $1\frac{1}{3} - \frac{1}{4}$

Estimate: _____

Problem Solving  REAL WORLD

13. For a fruit salad recipe, Jenna combined $\frac{3}{8}$ cup of raisins, $\frac{7}{8}$ cup of oranges, and $\frac{3}{4}$ cup of apples. About how many cups of fruit are in the salad?

_____

14. Tyler had $2\frac{7}{16}$ yards of fabric. He used $\frac{3}{4}$ yard to make a vest. About how much fabric did he have left?

_____

Lesson Check

1. Helen's house is located on a rectangular lot that is $1\frac{1}{8}$ miles by $\frac{9}{10}$ mile. Estimate the distance around the lot.

   Ⓐ about 3 miles

   Ⓑ about 4 miles

   Ⓒ about 5 miles

   Ⓓ about 6 miles

2. Keith bought a package with $2\frac{9}{16}$ pounds of ground meat to make hamburgers. He has $\frac{2}{5}$ pound of ground meat left. About how many pounds of ground meat did he use for the hamburgers?

   Ⓐ about 4 pounds

   Ⓑ about 3 pounds

   Ⓒ about 2 pounds

   Ⓓ about 1 pound

Spiral Review

3. Jason bought two identical boxes of nails. One box weighs 168 ounces. What is the total weight in ounces of the nails Jason bought? (Lesson 1.6)

   Ⓐ 84 ounces

   Ⓑ 226 ounces

   Ⓒ 326 ounces

   Ⓓ 336 ounces

4. Hank wants to divide 345 pieces of candy evenly among his 23 classmates. How many pieces will be left over? (Lesson 2.7)

   Ⓐ 0

   Ⓑ 2

   Ⓒ 11

   Ⓓ 22

5. Which is the most reasonable estimate for $23.63 \div 6$? (Lesson 5.3)

   Ⓐ 3

   Ⓑ 4

   Ⓒ 5

   Ⓓ 6

6. What is a rule for the sequence below?
   (Lesson 3.10)

   0.8, 0.86, 0.92, 0.98, ...

   Ⓐ start at 0.8, add 0.06

   Ⓑ start at 0.8, add 0.6

   Ⓒ start at 0.98, subtract 0.06

   Ⓓ start at 0.98, subtract 0.6

Name _____

Common Denominators and Equivalent Fractions

Use a common denominator to write an equivalent fraction for each fraction.

1. $\frac{1}{5}, \frac{1}{2}$ common denominator: __**10**__

Think: 10 is a multiple of 5 and 2. Find equivalent fractions with a denominator of 10.

_____

2. $\frac{1}{4}, \frac{2}{3}$ common denominator: _____

_____

3. $\frac{5}{6}, \frac{1}{3}$ common denominator: _____

_____

4. $\frac{3}{5}, \frac{1}{3}$ common denominator: _____

_____

5. $\frac{1}{2}, \frac{3}{8}$ common denominator: _____

_____

6. $\frac{1}{6}, \frac{1}{4}$ common denominator: _____

_____

Use the least common denominator to write an equivalent fraction for each fraction.

7. $\frac{5}{6}, \frac{2}{9}$

_____

8. $\frac{1}{12}, \frac{3}{8}$

_____

9. $\frac{5}{9}, \frac{2}{15}$

_____

Problem Solving  REAL WORLD

10. Ella spends $\frac{2}{3}$ hour practicing the piano each day. She also spends $\frac{1}{2}$ hour jogging. What is the least common denominator of the fractions?

_____

11. In a science experiment, a plant grew $\frac{3}{4}$ inch one week and $\frac{1}{2}$ inch the next week. Use a common denominator to write an equivalent fraction for each fraction.

_____

Lesson Check

1. Which fractions use the least common denominator and are equivalent to $\frac{9}{10}$ and $\frac{5}{6}$?

   (A) $\frac{54}{60}$ and $\frac{45}{60}$

   (B) $\frac{27}{30}$ and $\frac{25}{30}$

   (C) $\frac{29}{30}$ and $\frac{15}{30}$

   (D) $\frac{9}{16}$ and $\frac{5}{16}$

2. Joseph says that there is $\frac{5}{8}$ of a pumpkin pie left and $\frac{1}{2}$ of a peach pie left. Which is NOT a pair of equivalent fractions for $\frac{5}{8}$ and $\frac{1}{2}$?

   (A) $\frac{5}{8}$ and $\frac{4}{8}$

   (B) $\frac{10}{16}$ and $\frac{8}{16}$

   (C) $\frac{15}{24}$ and $\frac{8}{24}$

   (D) $\frac{50}{80}$ and $\frac{40}{80}$

Spiral Review

3. Matthew had the following times in two races: 3.032 minutes and 3.023 minutes. Which sentence about these two numbers is true? (Lesson 3.3)

   (A) 3.032 > 3.023

   (B) 3.032 = 3.023

   (C) 3.032 < 3.023

   (D) 3.023 > 3.023

4. Olivia's class collected 3,591 bottle caps in 57 days. On average, how many bottle caps did the class collect per day? (Lesson 2.6)

   (A) 57

   (B) 62

   (C) 63

   (D) 64

5. Elizabeth multiplied 0.63 by 1.8. Which is the correct product? (Lesson 4.7)

   (A) 0.567

   (B) 0.654

   (C) 1.114

   (D) 1.134

6. What is the value of $(17 + 8) - 6 \times 2$? (Lesson 1.11)

   (A) 13

   (B) 21

   (C) 37

   (D) 38

Add and Subtract Fractions

Find the sum or difference. Write your answer in simplest form.

1. $\frac{1}{2} - \frac{1}{7}$

$$\frac{1}{2} \rightarrow \frac{7}{14}$$
$$-\frac{1}{7} \rightarrow -\frac{2}{14}$$
$$\frac{5}{14}$$

2. $\frac{7}{10} - \frac{1}{2}$

_____

3. $\frac{1}{6} + \frac{1}{2}$

_____

4. $\frac{5}{8} + \frac{2}{5}$

_____

5. $\frac{9}{10} - \frac{1}{3}$

_____

6. $\frac{3}{4} - \frac{2}{5}$

_____

7. $\frac{5}{7} - \frac{1}{4}$

_____

8. $\frac{7}{8} + \frac{1}{3}$

_____

9. $\frac{5}{6} + \frac{2}{5}$

_____

10. $\frac{1}{6} - \frac{1}{10}$

_____

11. $\frac{6}{11} - \frac{1}{2}$

_____

12. $\frac{5}{6} + \frac{3}{7}$

_____

Problem Solving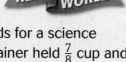

13. Kaylin mixed two liquids for a science experiment. One container held $\frac{7}{8}$ cup and the other held $\frac{9}{10}$ cup. What is the total amount of the mixture?

_____

14. Henry bought $\frac{1}{4}$ pound of screws and $\frac{2}{5}$ pound of nails to build a skateboard ramp. What is the total weight of the screws and nails?

_____

Lesson Check

1. Lyle bought $\frac{3}{8}$ pound of red grapes and $\frac{5}{12}$ pound of green grapes. How many pounds of grapes did he buy?

- (A) $\frac{19}{24}$ pound
- (B) $\frac{2}{5}$ pound
- (C) $\frac{1}{3}$ pound
- (D) $\frac{1}{24}$ pound

2. Jennifer had a $\frac{7}{8}$-foot board. She cut off a $\frac{1}{4}$-foot piece that was for a project. In feet, how much of the board was left?

- (A) $\frac{12}{8}$ feet
- (B) $\frac{9}{8}$ feet
- (C) $\frac{6}{8}$ foot
- (D) $\frac{5}{8}$ foot

Spiral Review

3. Ivan has 15 yards of green felt and 12 yards of blue felt to make 3 quilts. If Ivan uses the same total number of yards for each quilt, how many yards does he use for each quilt? **(Lesson 1.9)**

- (A) 4 yards
- (B) 5 yards
- (C) 9 yards
- (D) 27 yards

4. Eight identical shirts cost a total of $152. How much does one shirt cost? **(Lesson 2.2)**

- (A) $2
- (B) $8
- (C) $19
- (D) $24

5. Melissa bought a pencil for $0.34, an eraser for $0.22, and a notebook for $0.98. Which is the most reasonable estimate for the amount Melissa spent? **(Lesson 3.7)**

- (A) $1.60
- (B) $1.50
- (C) $1.40
- (D) $1.30

6. The 12 members in Dante's hiking club shared 176 ounces of trail mix equally. How many ounces of trail mix did each member receive? **(Lesson 2.7)**

- (A) 15 ounces
- (B) $14\frac{2}{3}$ ounces
- (C) 14 ounces
- (D) 12 ounces

Name _____

Add and Subtract Mixed Numbers

Find the sum or difference. Write your answer in simplest form.

1. $3\frac{1}{2} - 1\frac{1}{5}$

$$3\frac{1}{2} \rightarrow 3\frac{5}{10}$$
$$-1\frac{1}{5} \rightarrow -1\frac{2}{10}$$
$$\overline{\phantom{-1\frac{1}{5}} 2\frac{3}{10}}$$

2. $2\frac{1}{3} + 1\frac{3}{4}$

_____

3. $4\frac{1}{8} + 2\frac{1}{3}$

_____

4. $5\frac{1}{3} + 6\frac{1}{6}$

_____

5. $2\frac{1}{4} + 1\frac{2}{5}$

_____

6. $5\frac{17}{18} - 2\frac{2}{3}$

_____

7. $6\frac{3}{4} - 1\frac{5}{8}$

_____

8. $5\frac{3}{7} - 2\frac{1}{5}$

_____

9. $4\frac{1}{8} + 2\frac{5}{12}$

_____

10. $6\frac{6}{7} - 2\frac{3}{4}$

_____

11. $5\frac{5}{6} - 2\frac{3}{4}$

_____

12. $2\frac{6}{25} - 1\frac{1}{10}$

_____

Problem Solving REAL WORLD

13. Jacobi bought $7\frac{1}{2}$ pounds of meatballs. He decided to cook $1\frac{1}{4}$ pounds and freeze the rest. How many pounds did he freeze?

_____

14. Jill walked $8\frac{1}{8}$ miles to a park and then $7\frac{2}{5}$ miles home. How many miles did she walk in all?

_____

Lesson Check

1. Ming has a goal to jog $4\frac{1}{2}$ miles each day. On Monday she jogged $5\frac{9}{16}$ miles. By how much did she exceed her goal for that day?

   (A) $1\frac{1}{16}$ miles

   (B) $1\frac{7}{16}$ miles

   (C) $1\frac{8}{16}$ miles

   (D) $1\frac{8}{14}$ miles

2. At the deli, Ricardo ordered $3\frac{1}{5}$ pounds of cheddar cheese and $2\frac{3}{4}$ pounds of mozzarella cheese. How many pounds of cheese did he order?

   (A) $5\frac{19}{20}$ pounds

   (B) $5\frac{17}{20}$ pounds

   (C) $5\frac{4}{9}$ pounds

   (D) $5\frac{4}{20}$ pounds

Spiral Review

3. The theater has 175 seats. There are 7 seats in each row. How many rows are there?
   (Lesson 2.2)

   (A) 15

   (B) 17

   (C) 25

   (D) 30

4. Over the first 14 days, 2,755 people visted a new store. About how many people visited the store each day? (Lesson 2.5)

   (A) about 100

   (B) about 150

   (C) about 200

   (D) about 700

5. Which number is 100 times as great as 0.3?
   (Lesson 3.2)

   (A) 300

   (B) 30

   (C) 3

   (D) 0.003

6. Mark said that the product of 0.02 and 0.7 is 14. Mark is wrong. Which product is correct? (Lesson 4.8)

   (A) 0.014

   (B) 0.14

   (C) 1.4

   (D) 14.0

Name _____

Subtraction with Renaming

Estimate. Then find the difference and write it in simplest form.

1. Estimate: _____

$$6\frac{1}{3} - 1\frac{2}{5}$$

$$6\frac{1}{3} \rightarrow 6\frac{\overset{20}{\cancel{5}}}{15}$$

$$-1\frac{2}{5} \rightarrow -1\frac{6}{15}$$

$$\rule{3cm}{0.4pt}$$

$$4\frac{14}{15}$$

2. Estimate: _____

$$4\frac{1}{2} - 3\frac{5}{6}$$

3. Estimate: _____

$$9 - 3\frac{7}{8}$$

4. Estimate: _____

$$2\frac{1}{6} - 1\frac{2}{7}$$

5. Estimate: _____

$$8 - 6\frac{1}{9}$$

6. Estimate: _____

$$9\frac{1}{4} - 3\frac{2}{3}$$

7. Estimate: _____

$$2\frac{1}{8} - 1\frac{2}{7}$$

8. Estimate: _____

$$8\frac{1}{5} - 3\frac{5}{9}$$

9. Estimate: _____

$$10\frac{2}{3} - 5\frac{9}{10}$$

Problem Solving REAL WORLD

10. Carlene bought $8\frac{1}{16}$ yards of ribbon to decorate a shirt. She only used $5\frac{1}{2}$ yards. How much ribbon does she have left over?

11. During his first vet visit, Pedro's puppy weighed $6\frac{1}{8}$ pounds. On his second visit, he weighed $9\frac{1}{16}$ pounds. How much weight did he gain between visits?

_____

_____

Lesson Check

1. Natalia picked $7\frac{1}{6}$ bushels of apples today and $4\frac{5}{8}$ bushels yesterday. How many more bushels did she pick today?

   (A) $3\frac{4}{24}$ bushels   (C) $2\frac{4}{8}$ bushels

   (B) $2\frac{13}{24}$ bushels   (D) $1\frac{6}{12}$ bushels

2. Max needs $10\frac{1}{4}$ cups flour to make a batch of pizza dough for the pizzeria. He only has $4\frac{1}{2}$ cups flour. How much more flour does he need to make the dough?

   (A) $6\frac{1}{4}$ cups   (C) $5\frac{1}{2}$ cups

   (B) $5\frac{3}{4}$ cups   (D) $5\frac{1}{4}$ cups

Spiral Review

3. The accountant charged $35 for the first hour of work and $23 for each hour after that. He earned a total of $127. How many hours did he work? (Lesson 1.9)

   (A) 2 hours

   (B) 3 hours

   (C) 4 hours

   (D) 5 hours

4. The soccer league needs to transport all 133 players to the tournament. If 4 players can ride in one car, how many cars are needed? (Lesson 2.2)

   (A) 25

   (B) 30

   (C) 33

   (D) 34

5. Which number shows five hundred million, one hundred fifteen in standard form?
   (Lesson 1.2)

   (A)   5,115,000

   (B)   5,000,115

   (C) 500,115,000

   (D) 500,000,115

6. Find the quotient. (Lesson 5.6)

   $$6.39 \div 0.3$$

   (A)   0.213

   (B)   2.13

   (C)   21.3

   (D) 213.0

Name _____

Patterns with Fractions

Write a rule for the sequence. Then, find the unknown term.

1. $\frac{1}{2}, \frac{2}{3}, \underline{\frac{5}{6}}, 1, 1\frac{1}{6}$

Think: The pattern is increasing.
Add $\frac{1}{6}$ to find the next term.

2. $1\frac{3}{8}, 1\frac{3}{4}, 2\frac{1}{8}, \underline{\hspace{1cm}}, 2\frac{7}{8}$

Rule: _____

Rule: _____

3. $1\frac{9}{10}, 1\frac{7}{10}, \underline{\hspace{1cm}}, 1\frac{3}{10}, 1\frac{1}{10}$

4. $2\frac{5}{12}, 2\frac{1}{6}, 1\frac{11}{12}, \underline{\hspace{1cm}}, 1\frac{5}{12}$

Rule: _____

Rule: _____

Write the first four terms of the sequence.

5. **Rule:** start at $\frac{1}{2}$, add $\frac{1}{3}$

6. **Rule:** start at $3\frac{1}{8}$, subtract $\frac{3}{4}$

_____

_____

7. **Rule:** start at $5\frac{1}{2}$, add $1\frac{1}{5}$

8. **Rule:** start at $6\frac{2}{3}$, subtract $1\frac{1}{4}$

_____

_____

Problem Solving REAL WORLD

9. Jarett's puppy weighed $3\frac{3}{4}$ ounces at birth. At one week old, the puppy weighed $5\frac{1}{8}$ ounces. At two weeks old, the puppy weighed $6\frac{1}{2}$ ounces. If the weight gain continues in this pattern, how much will the puppy weigh at three weeks old?

10. A baker started out with 12 cups of flour. She had $9\frac{1}{4}$ cups of flour left after the first batch of batter she made. She had $6\frac{1}{2}$ cups of flour left after the second batch of batter she made. If she makes two more batches of batter, how many cups of flour will be left?

_____

_____

Lesson Check

1. What is a rule for the sequence?

   $\frac{5}{6}, 1\frac{1}{2}, 2\frac{1}{6}, 2\frac{5}{6}, \ldots$

   (A) add $1\frac{1}{4}$

   (B) add $\frac{2}{3}$

   (C) subtract $1\frac{1}{4}$

   (D) subtract $\frac{2}{3}$

2. Jaime biked $5\frac{1}{4}$ miles on Monday, $6\frac{7}{8}$ miles on Tuesday, and $8\frac{1}{2}$ miles on Wednesday. If he continues the pattern, how many miles will he bike on Friday?

   (A) $10\frac{1}{8}$ miles

   (B) $10\frac{3}{4}$ miles

   (C) $11\frac{1}{8}$ miles

   (D) $11\frac{3}{4}$ miles

Spiral Review

3. Jaylyn rode her bicycle in a bike-a-thon. She rode 33.48 miles in 2.7 hours. If she rode at the same speed, what was her speed in miles per hour? **(Lesson 5.6)**

   (A) 12.04

   (B) 12.08

   (C) 12.4

   (D) 12.8

4. One week a company filled 546 boxes with widgets. Each box held 38 widgets. How many widgets did the company pack in boxes that week? **(Lesson 1.7)**

   (A) 20,748

   (B) 20,608

   (C) 6,006

   (D) 2,748

5. Which expression represents the statement "Add 9 and 3, then multiply by 6"? **(Lesson 1.10)**

   (A) $9 + 3 \times 6$

   (B) $6 \times (9 + 3)$

   (C) $6 \times 9 + 3$

   (D) $6 \times 9 \times 3$

6. Mason took 9.4 minutes to complete the first challenge in the Champs Challenge. He completed the second challenge 2.65 minutes faster than the first challenge. How long did it take Mason to complete the second challenge? **(Lesson 3.9)**

   (A) 7.39 minutes

   (B) 7.35 minutes

   (C) 6.85 minutes

   (D) 6.75 minutes

Name _____

Problem Solving • Practice Addition and Subtraction

Read each problem and solve.

1. From a board 8 feet in length, Emmet cut two $2\frac{1}{3}$-foot bookshelves. How much of the board remained?

   Write an equation: $8 = 2\frac{1}{3} + 2\frac{1}{3} + x$

   Rewrite the equation to work backward:

   $$8 - 2\frac{1}{3} - 2\frac{1}{3} = x$$

   Subtract twice to find the length remaining: $3\frac{1}{3}$ **feet**

2. Lynne bought a bag of grapefruit, $1\frac{5}{8}$ pounds of apples, and $2\frac{3}{16}$ pounds of bananas. The total weight of her purchases was $7\frac{1}{2}$ pounds. How much did the bag of grapefruit weigh?

   _____

3. Mattie's house consists of two stories and an attic. The first floor is $8\frac{5}{6}$ feet tall, the second floor is $8\frac{1}{2}$ feet tall, and the entire house is $24\frac{1}{3}$ feet tall. How tall is the attic?

   _____

4. It is $10\frac{3}{5}$ miles from Alston to Barton and $12\frac{1}{2}$ miles from Barton to Chester. The distance from Alston to Durbin, via Barton and Chester, is 35 miles. How far is it from Chester to Durbin?

   _____

5. Marcie bought a 50-foot roll of packing tape. She used two $8\frac{5}{6}$-foot lengths. How much tape is left on the roll?

   _____

6. Meg started her trip with $11\frac{1}{2}$ gallons of gas in her car's gas tank. She bought an additional $6\frac{4}{5}$ gallons on her trip and arrived back home with $3\frac{3}{10}$ gallons left. How much gas did she use on the trip?

   _____

Lesson Check

1. Paula spent $\frac{3}{8}$ of her allowance on clothes and $\frac{1}{6}$ on entertainment. What fraction of her allowance did she spend on other items?

   (A) $\frac{3}{8}$

   (B) $\frac{11}{24}$

   (C) $\frac{13}{24}$

   (D) $\frac{5}{8}$

2. Della bought a tree seedling that was $2\frac{1}{4}$ feet tall. During the first year, it grew $1\frac{1}{6}$ feet. After two years, it was 5 feet tall. How much did the seedling grow during the second year?

   (A) $1\frac{1}{4}$ feet

   (B) $1\frac{1}{3}$ feet

   (C) $1\frac{5}{12}$ feet

   (D) $1\frac{7}{12}$ feet

Spiral Review

3. Which is another way to write 100,000? (Lesson 1.4)

   (A) 10^6

   (B) 10^5

   (C) 10×10^5

   (D) 10×10^6

4. Which expression is the best choice for estimating $868 \div 28$? (Lesson 2.5)

   (A) $868 \div 28$

   (B) $900 \div 30$

   (C) $1{,}000 \div 20$

   (D) $1{,}000 \div 30$

5. Justin gave the clerk $20 to pay a bill of $6.57. How much change should Justin get? (Lesson 3.11)

   (A) $12.43

   (B) $12.53

   (C) $13.43

   (D) $14.43

6. What is the value of the following expression?
   $$7 + 18 \div (6 - 3)$$ (Lesson 1.12)

   (A) 9

   (B) 13

   (C) 21

   (D) 27

Name _____

Use Properties of Addition

Use the properties and mental math to solve. Write your answer in simplest form.

1. $\left(2\frac{1}{3} + 1\frac{2}{5}\right) + 3\frac{2}{3}$

$= \left(1\frac{2}{5} + 2\frac{1}{3}\right) + 3\frac{2}{3}$

$= 1\frac{2}{5} + \left(2\frac{1}{3} + 3\frac{2}{3}\right)$

$= 1\frac{2}{5} + 6$

$= 7\frac{2}{5}$

2. $8\frac{1}{5} + \left(4\frac{2}{5} + 3\frac{3}{10}\right)$

3. $\left(1\frac{3}{4} + 2\frac{3}{8}\right) + 5\frac{7}{8}$

4. $2\frac{1}{10} + \left(1\frac{2}{7} + 4\frac{9}{10}\right)$

5. $\left(4\frac{3}{5} + 6\frac{1}{3}\right) + 2\frac{3}{5}$

6. $1\frac{1}{4} + \left(3\frac{2}{3} + 5\frac{3}{4}\right)$

7. $\left(7\frac{1}{8} + 1\frac{2}{7}\right) + 4\frac{3}{7}$

8. $3\frac{1}{4} + \left(3\frac{1}{4} + 5\frac{1}{5}\right)$

9. $6\frac{2}{3} + \left(5\frac{7}{8} + 2\frac{1}{3}\right)$

Problem Solving REAL WORLD

10. Elizabeth rode her bike $6\frac{1}{2}$ miles from her house to the library and then another $2\frac{2}{5}$ miles to her friend Milo's house. If Carson's house is $2\frac{1}{2}$ miles beyond Milo's house, how far would she travel from her house to Carson's house?

11. Hassan made a vegetable salad with $2\frac{3}{8}$ pounds of tomatoes, $1\frac{1}{4}$ pounds of asparagus, and $2\frac{7}{8}$ pounds of potatoes. How many pounds of vegetables did he use altogether?

Lesson Check

1. What is the sum of $2\frac{1}{3}$, $3\frac{5}{6}$, and $6\frac{2}{3}$?

   (A) $12\frac{5}{6}$

   (B) $11\frac{5}{6}$

   (C) $11\frac{8}{12}$

   (D) $11\frac{10}{18}$

2. Letitia has $7\frac{1}{6}$ yards of yellow ribbon, $5\frac{1}{4}$ yards of orange ribbon, and $5\frac{1}{6}$ yards of brown ribbon. How much ribbon does she have altogether?

   (A) $18\frac{7}{12}$ yards

   (B) $18\frac{1}{6}$ yards

   (C) $17\frac{7}{12}$ yards

   (D) $17\frac{3}{16}$ yards

Spiral Review

3. Juanita wrote 3×47 as $3 \times 40 + 3 \times 7$. Which property did she use to rewrite the expression? (Lesson 1.3)

   (A) Associative Property of Multiplication

   (B) Commutative Property of Multiplication

   (C) Distributive Property

   (D) Identity Property

4. What is the value of the expression $18 - 2 \times (4 + 3)$. (Lesson 1.11)

   (A) 4

   (B) 7

   (C) 13

   (D) 112

5. Evan spent $15.89 on 7 pounds of birdseed. How much did the birdseed cost per pound?
   (Lesson 5.4)

   (A) $2.07

   (B) $2.12

   (C) $2.27

   (D) $2.29

6. Cade rode $1\frac{3}{5}$ miles on Saturday and $1\frac{3}{4}$ miles on Sunday. How far did he ride in all on the two days? (Lesson 6.6)

   (A) $2\frac{7}{20}$ miles

   (B) $2\frac{9}{20}$ miles

   (C) $3\frac{3}{10}$ miles

   (D) $3\frac{7}{20}$ miles

Name _____

Chapter 6 Extra Practice

Lessons 6.1 - 6.2

Use fraction strips to find the sum or difference. Write your answer in simplest form.

1. $\frac{5}{8} + \frac{1}{4}$ **2.** $\frac{7}{10} - \frac{3}{5}$ **3.** $\frac{1}{9} + \frac{5}{6}$ **4.** $\frac{3}{4} - \frac{5}{8}$

Lesson 6.3

Estimate the sum or difference.

1. $\frac{6}{10} + \frac{7}{12}$ **2.** $\frac{5}{12} + \frac{7}{8}$ **3.** $1\frac{3}{8} - \frac{8}{9}$

Lesson 6.4

Use a common denominator to write an equivalent fraction for each fraction.

1. $\frac{1}{2}, \frac{1}{3}$

Common
denominator: _____

2. $\frac{7}{8}, \frac{3}{10}$

Common
denominator: _____

3. $\frac{2}{3}, \frac{3}{4}$

Common
denominator: _____

_____   _____   _____

Use the least common denominator to write an equivalent fraction for each fraction.

4. $\frac{1}{4}, \frac{5}{6}$ **5.** $\frac{1}{2}, \frac{1}{8}$ **6.** $\frac{3}{5}, \frac{2}{7}$

_____   _____   _____

Lessons 6.5 – 6.7

Find the sum or difference. Write your answer in simplest form.

1. $\dfrac{7}{8} - \dfrac{5}{6}$

2. $5 - 2\dfrac{4}{5}$

3. $3\dfrac{1}{4} + 1\dfrac{7}{8}$

4. $6\dfrac{9}{10} - 5\dfrac{4}{5}$

5. $\dfrac{1}{3} + \dfrac{4}{15}$

6. $1\dfrac{1}{3} + \dfrac{2}{5}$

7. $2\dfrac{3}{8} + 8\dfrac{5}{6}$

8. $9\dfrac{1}{4} - 2\dfrac{5}{8}$

Lesson 6.8

1. On the first day of the play, the auditorium was $\dfrac{1}{3}$ full, the second day it was $\dfrac{5}{12}$ full, and on the third day it was $\dfrac{1}{2}$ full. If this pattern continues, how full will it be on the fourth day?

2. Jake set up a study schedule. The plan called for him to study $\dfrac{1}{4}$ hour, $\dfrac{5}{8}$ hour, and 1 hour on Monday, Tuesday, and Wednesday in that order. If he continues with this pattern, how long will he study on Friday?

Lesson 6.9

1. Sierra spent $\dfrac{2}{3}$ of her earnings on clothes and $\dfrac{1}{5}$ on school supplies. She saved the rest. What fraction of her earnings did she save?

2. Noah made $1\dfrac{1}{2}$ dozen blueberry muffins and $1\dfrac{3}{4}$ dozen lemon muffins. He needs to take 5 dozen muffins to the bake sale. How many dozen more muffins does he need to bake?

Lesson 6.10

Use the properties and mental math to solve. Write your answer in simplest form.

1. $\left(\dfrac{4}{5} + \dfrac{2}{3}\right) + \dfrac{1}{5}$

2. $1\dfrac{1}{4} + \left(\dfrac{3}{4} + \dfrac{2}{7}\right)$

3. $\left(\dfrac{1}{6} + \dfrac{4}{5}\right) + \dfrac{5}{6}$

School-Home Letter

Dear Family,

Throughout the next few weeks, our math class will be learning about multiplying fractions and mixed numbers. We will also be using area models to help understand fraction multiplication.

You can expect to see homework with real-world problems that involve multiplication with fractions and mixed numbers.

Here is a sample of how your child is taught to multiply two mixed numbers.

Vocabulary

denominator The part of the fraction below the line, which tells how many equal parts there are in the whole or in a group

mixed number A number represented by a whole number and a fraction

numerator The part of a fraction above the line, which tells how many parts are being counted

product The answer in a multiplication problem

simplest form A fraction in which 1 is the only number that can divide evenly into the numerator and the denominator

🔒 MODEL Multiply Mixed Numbers

Multipy. $1\frac{3}{4} \times 2\frac{1}{2}$

STEP 1

Write the mixed numbers as fractions.

STEP 2

Multiply the fractions.

STEP 3

Write the product as a mixed number in simplest form.

$$1\frac{3}{4} \times 2\frac{1}{2} = \frac{7}{4} \times \frac{5}{2}$$

$$= \frac{7 \times 5}{4 \times 2}$$

$$= \frac{35}{8}$$

$$= 4\frac{3}{8}$$

Tips
Checking for Reasonable Answers

When a fraction is multiplied by 1, the product equals the fraction. When a fraction is multiplied by a factor greater than 1, the product will be greater than the fraction. When a fraction is multiplied by a factor less than 1, the product will be less than either factor.

Activity

Use recipes to practice multiplication with fractions and mixed numbers. Work together to solve problems such as, "One batch of the recipe calls for $2\frac{1}{4}$ cups of flour. How much flour would we need to make $1\frac{1}{2}$ batches?"

Carta para la casa

Vocabulario

denominador La parte de la fracción debajo de la barra que indica cuántas partes iguales hay en un total o en un grupo

número mixto Un número representado por un número entero y una fracción

numerador La parte de una fracción por encima de la barra que indica cuántas partes se están contando

producto El resultado de un problema de multiplicación

mínima expresión Una fracción en la que 1 es el único número que se puede dividir en partes iguales entre el numerador y el denominador

Querida familia,

Durante las próximas semanas, en la clase de matemáticas aprenderemos a multiplicar fracciones y números mixtos. También aprenderemos a usar modelos de área para entender la multiplicación de fracciones.

Llevaré a la casa tareas con problemas del mundo real que involucren la multiplicación con fracciones y números mixtos.

Este es un ejemplo de la manera como aprenderemos a multiplicar dos números mixtos.

🔑 MODELO   Multiplicar números mixtos

Multiplica. $1\frac{3}{4} \times 2\frac{1}{2}$

PASO 1

Escribe los números mixtos como fracciones.

PASO 2

Multiplica las fracciones.

PASO 3

Escribe el producto como un número mixto en su mínima expresión.

$$1\frac{3}{4} \times 2\frac{1}{2} = \frac{7}{4} \times \frac{5}{2}$$

$$= \frac{7 \times 5}{4 \times 2}$$

$$= \frac{35}{8}$$

$$= 4\frac{3}{8}$$

Pistas

Buscar respuestas razonables

Cuando una fracción se multiplica por 1, el producto es igual a la fracción. Cuando una fracción se multiplica por un factor mayor que 1, el producto será mayor que la fracción. Cuando una fracción se multiplica por un factor menor que 1, el producto será menor que ese factor.

Actividad

Usen recetas para practicar la multiplicación con fracciones y números mixtos. Trabajen juntos para resolver problemas como, "Una porción de la receta pide $2\frac{1}{4}$ tazas de harina. ¿Cuánta harina necesitaremos para hacer $1\frac{1}{2}$ porciones?"

Name _____

Find Part of a Group

Use a model to solve.

1. $\frac{3}{4} \times 12 =$ __**9**__

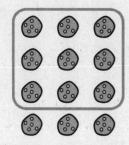

2. $\frac{7}{8} \times 16 =$ _____

3. $\frac{6}{10} \times 10 =$ _____

4. $\frac{2}{3} \times 9 =$ _____

5. $\frac{1}{6} \times 18 =$ _____

6. $\frac{4}{5} \times 10 =$ _____

Problem Solving REAL WORLD

7. Marco drew 20 pictures. He drew $\frac{3}{4}$ of them in art class. How many pictures did Marco draw in art class?

_____

8. Caroline has 10 marbles. One half of them are blue. How many of Caroline's marbles are blue?

_____

Lesson Check

1. Use the model to find $\frac{1}{3} \times 15$.

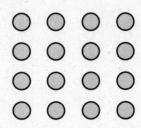

   (A) 3   (C) 6
   (B) 5   (D) 10

2. Use the model to find $\frac{2}{4} \times 16$.

   (A) 4   (C) 8
   (B) 6   (D) 12

Spiral Review

3. What is the value of the underlined digit? (Lesson 1.2)

   $\underline{6},560$

   (A) 6,000
   (B) 600
   (C) 60
   (D) 6

4. Nigel has 138 ounces of lemonade. How many 6-ounce servings of lemonade can he make? (Lesson 2.2)

   (A) 828
   (B) 132
   (C) 33
   (D) 23

5. Rafi had a board that was $15\frac{1}{2}$ feet long. He cut three pieces off the board that are each $3\frac{7}{8}$ feet long. How much of the board is left? (Lesson 6.6)

   (A) $3\frac{7}{8}$ feet
   (B) $7\frac{3}{4}$ feet
   (C) $11\frac{5}{8}$ feet
   (D) $13\frac{9}{16}$ feet

6. Susie spent $4\frac{1}{4}$ hours on Monday and $3\frac{5}{8}$ hours on Tuesday working on a history project. About how long did she spend working on the project? (Lesson 6.3)

   (A) 1 hour
   (B) 7 hours
   (C) 8 hours
   (D) 9 hours

Name _____

Multiply Fractions and Whole Numbers

Use the model to find the product.

1. $\frac{5}{12} \times 3 = \underline{\dfrac{5}{4}}$, or $1\frac{1}{4}$

1	1	1
$\frac{1}{4}$ $\frac{1}{4}$ $\frac{1}{4}$ $\frac{1}{4}$ $\frac{1}{4}$	$\frac{1}{4}$ $\frac{1}{4}$ $\frac{1}{4}$	$\frac{1}{4}$ $\frac{1}{4}$ $\frac{1}{4}$ $\frac{1}{4}$

2. $3 \times \frac{3}{4} =$ _____

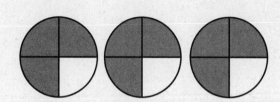

Find the product.

3. $\frac{2}{5} \times 5 =$ _____

4. $7 \times \frac{2}{3} =$ _____

5. $\frac{3}{8} \times 4 =$ _____

6. $7 \times \frac{5}{6} =$ _____

7. $\frac{5}{12} \times 6 =$ _____

8. $9 \times \frac{2}{3} =$ _____

Problem Solving REAL WORLD

9. Jody has a 5-pound bag of potatoes. She uses $\frac{4}{5}$ of the bag to make potato salad. How many pounds of potatoes does Jody use for the potato salad?

10. Lucas lives $\frac{5}{8}$ mile from school. Kenny lives twice as far as Lucas from school. How many miles does Kenny live from school?

_____

_____

Lesson Check

1. In gym class, Ted runs $\frac{4}{5}$ mile. His teacher runs 6 times that distance each day. How many miles does Ted's teacher run each day?

   Ⓐ $\frac{5}{24}$ mile

   Ⓑ $3\frac{1}{3}$ miles

   Ⓑ $4\frac{4}{5}$ miles

   Ⓓ $7\frac{1}{2}$ miles

2. Jon is decorating a banner for a parade. Jon uses a piece of red ribbon, which is $\frac{3}{4}$ yard long. Jon also needs blue ribbon that is 5 times as long as the red ribbon. How much blue ribbon does Jon need?

   Ⓐ $\frac{3}{20}$ yard

   Ⓑ $3\frac{3}{4}$ yards

   Ⓒ $4\frac{1}{4}$ yards

   Ⓓ $7\frac{1}{3}$ yards

Spiral Review

3. Mirror Lake Elementary School has 168 students and chaperones going on the fifth grade class trip. Each bus can hold 54 people. What is the least number of buses needed for the trip? **(Lesson 2.7)**

   Ⓐ 3

   Ⓑ 4

   Ⓒ 5

   Ⓓ 8

4. From an 8-foot board, a carpenter sawed off one piece that was $2\frac{3}{4}$ feet long and another piece that was $3\frac{1}{2}$ feet long. How much of the board was left? **(Lesson 6.9)**

   Ⓐ $1\frac{3}{4}$ feet

   Ⓑ $2\frac{1}{4}$ feet

   Ⓒ $2\frac{3}{4}$ feet

   Ⓓ $6\frac{1}{4}$ feet

5. Which expression does NOT have a value of 18? **(Lesson 1.11)**

   Ⓐ $8 \div 4 \times (3 + 6)$

   Ⓑ $(20 - 13) \times 4 - 10$

   Ⓒ $9 + 3 \times 5 - 6$

   Ⓓ $30 - 5 \times 4 + 2$

6. Which of the following decimals has the least value? **(Lesson 3.3)**

   Ⓐ 0.3

   Ⓑ 0.029

   Ⓒ 0.003

   Ⓓ 0.01

Name _____

Fraction and Whole Number Multiplication

Find the product. Write the product in simplest form.

1. $4 \times \frac{5}{8} =$ _____ $2\frac{1}{2}$

$4 \times \frac{5}{8} = \frac{20}{8}$

$\frac{20}{8} = 2\frac{4}{8}$, or $2\frac{1}{2}$

2. $\frac{2}{9} \times 3 =$ _____

3. $\frac{4}{5} \times 10 =$ _____

4. $\frac{3}{4} \times 9 =$ _____

5. $8 \times \frac{5}{6} =$ _____

6. $7 \times \frac{1}{2} =$ _____

7. $\frac{2}{5} \times 6 =$ _____

8. $9 \times \frac{2}{3} =$ _____

9. $\frac{3}{10} \times 9 =$ _____

10. $4 \times \frac{3}{8} =$ _____

11. $\frac{3}{5} \times 7 =$ _____

12. $\frac{1}{8} \times 6 =$ _____

Problem Solving 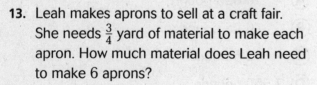 REAL WORLD

13. Leah makes aprons to sell at a craft fair. She needs $\frac{3}{4}$ yard of material to make each apron. How much material does Leah need to make 6 aprons?

14. The gas tank of Mr. Tanaka's car holds 15 gallons of gas. He used $\frac{2}{3}$ of a tank of gas last week. How many gallons of gas did Mr. Tanaka use?

_____   _____

Lesson Check

1. At the movies, Liz eats $\frac{1}{4}$ of a box of popcorn. Her friend Kyra eats two times as much popcorn as Liz eats. How much of a box of popcorn does Kyra eat?

   (A) $\frac{1}{16}$

   (B) $\frac{1}{8}$

   (C) $\frac{1}{4}$

   (D) $\frac{1}{2}$

2. It takes Ed 45 minutes to complete his science homework. It takes him $\frac{2}{3}$ as long to complete his math homework. How long does it take Ed to complete his math homework?

   (A) 15 minutes

   (B) 30 minutes

   (C) 90 minutes

   (D) 120 minutes

Spiral Review

3. Which is the best estimate for the quotient? **(Lesson 5.3)**

   $$591.3 \div 29$$

   (A) about 2

   (B) about 3

   (C) about 20

   (D) about 30

4. Sandy bought $\frac{3}{4}$ yard of red ribbon and $\frac{2}{3}$ yard of white ribbon to make some hair bows. Altogether, how many yards of ribbon did she buy? **(Lesson 6.5)**

   (A) $\frac{5}{12}$ yard

   (B) $\frac{5}{7}$ yard

   (C) $1\frac{5}{12}$ yards

   (D) $1\frac{7}{12}$ yards

5. Eric jogged $3\frac{1}{4}$ miles on Monday, $5\frac{5}{8}$ miles on Tuesday, and 8 miles on Wednesday. Suppose he continues the pattern for the remainder of the week. How far will Eric jog on Friday? **(Lesson 6.8)**

   (A) $10\frac{3}{8}$ miles

   (B) $10\frac{3}{4}$ miles

   (C) $12\frac{3}{8}$ miles

   (D) $12\frac{3}{4}$ miles

6. Sharon bought 25 pounds of ground beef and made 100 hamburger patties of equal weight. What is the weight of each hamburger patty? **(Lesson 5.1)**

   (A) 0.025 pound

   (B) 0.25 pound

   (C) 2.5 pounds

   (D) 2,500 pounds

Multiply Fractions

Find the product.

1.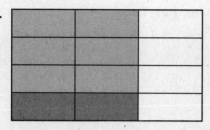

$\frac{1}{4} \times \frac{2}{3} = \underline{\frac{2}{12}}, \text{ or } \underline{\frac{1}{6}}$

2.

$\frac{2}{5} \times \frac{5}{6} = \underline{\hspace{3cm}}$

Find the product. Draw a model.

3. $\frac{4}{5} \times \frac{1}{2} = \underline{\hspace{3cm}}$

4. $\frac{3}{4} \times \frac{1}{3} = \underline{\hspace{3cm}}$

5. $\frac{3}{8} \times \frac{2}{3} = \underline{\hspace{3cm}}$

6. $\frac{3}{5} \times \frac{3}{5} = \underline{\hspace{3cm}}$

Problem Solving REAL WORLD

7. Nora has a piece of ribbon that is $\frac{3}{4}$ yard long. She will use $\frac{1}{2}$ of it to make a bow. What length of the ribbon will she use for the bow?

8. Marlon bought $\frac{7}{8}$ pound of turkey at the deli. He used $\frac{2}{3}$ of it to make sandwiches for lunch. How much of the turkey did Marlon use for sandwiches?

_____

_____

Lesson Check

1. Tina has $\frac{3}{5}$ pound of rice. She will use $\frac{2}{3}$ of it to make fried rice for her family. How much rice will Tina use to make fried rice?

   (A) $\frac{5}{8}$ pound

   (B) $\frac{3}{5}$ pound

   (C) $\frac{2}{5}$ pound

   (D) $\frac{1}{3}$ pound

2. The Waterfall Trail is $\frac{3}{4}$ mile long. At $\frac{1}{6}$ of the distance from the trailhead, there is a lookout. In miles, how far is the lookout from the trailhead?

   (A) $\frac{1}{8}$ mile

   (B) $\frac{1}{4}$ mile

   (C) $\frac{4}{10}$ mile

   (D) $\frac{24}{3}$ miles

Spiral Review

3. Hayden bought 48 new trading cards. Three-fourths of the new cards are baseball cards. How many baseball cards did Hayden buy? **(Lesson 7.1)**

   (A) 12

   (B) 16

   (C) 24

   (D) 36

4. Yesterday, Annie walked $\frac{9}{10}$ mile to her friend's house. Together, they walked $\frac{1}{3}$ mile to the library. Which is the best estimate for how far Annie walked yesterday? **(Lesson 6.3)**

   (A) about $\frac{1}{2}$ mile

   (B) about 1 mile

   (C) about $1\frac{1}{2}$ miles

   (D) about 2 miles

5. Erin is going to sew a jacket and a skirt. She needs $2\frac{3}{4}$ yards of material for the jacket and $1\frac{1}{2}$ yards of material for the skirt. Altogether, how many yards of material does Erin need? **(Lesson 6.6)**

   (A) $2\frac{3}{8}$ yards

   (B) $3\frac{1}{4}$ yards

   (C) $3\frac{7}{8}$ yards

   (D) $4\frac{1}{4}$ yards

6. Which of the following expressions simplifies to 4? **(Lesson 1.12)**

   (A) $[(3 \times 6) - (5 \times 2)] + 7$

   (B) $[(3 \times 6) + (5 \times 2)] \div 7$

   (C) $[(3 \times 6) + (5 + 2)] - 7$

   (D) $[(3 \times 6) - (5 \times 2)] \times 7$

Name _____

Compare Fraction Factors and Products

Complete the statement with *equal to, greater than,* or *less than.*

1. $\frac{3}{5} \times \frac{4}{7}$ will be ___**less than**___ $\frac{4}{7}$.

    Think: $\frac{4}{7}$ **is multiplied by a number less than 1;**
    so, $\frac{3}{5} \times \frac{4}{7}$ **will be less than** $\frac{4}{7}$**.**

2. $5 \times \frac{7}{8}$ will be _____ $\frac{7}{8}$.

3. $6 \times \frac{2}{5}$ will be _____ $\frac{2}{5}$.

4. $\frac{1}{9} \times 1$ will be _____ $\frac{1}{9}$.

5. $\frac{7}{8} \times \frac{3}{5}$ will be _____ $\frac{3}{5}$.

6. $\frac{4}{5} \times \frac{7}{7}$ will be _____ $\frac{4}{5}$.

Problem Solving REAL WORLD

7. Starla is making hot cocoa. She plans to multiply the recipe by 4 to make enough hot cocoa for the whole class. If the recipe calls for $\frac{1}{2}$ teaspoon vanilla extract, will she need more than $\frac{1}{2}$ teaspoon or less than $\frac{1}{2}$ teaspoon of vanilla extract to make all the hot cocoa?

_____

8. Miles is planning to spend $\frac{2}{3}$ as many hours bicycling this week as he did last week. Is Miles going to spend more hours or fewer hours bicycling this week than last week?

_____

Lesson Check

1. Trevor saves $\frac{2}{3}$ of the money he earns at his after-school job. Suppose Trevor starts saving $\frac{1}{4}$ as much as he is saving now. Which statement below will be true?

   (A) He will be saving four times as much.

   (B) He will be saving less.

   (C) He will be saving more.

   (D) He will be saving the same amount.

2. Suppose you multiply a whole number greater than 1 by the fraction $\frac{3}{5}$. Which statement below will be true?

   (A) The product will be equal to $\frac{3}{5}$.

   (B) The product will be greater than $\frac{3}{5}$.

   (C) The product will be less than $\frac{3}{5}$.

   (D) You cannot make any conclusions about the product.

Spiral Review

3. In the next 10 months, Colin wants to save $900 for his vacation. He plans to save $75 each of the first 8 months. How much must he save each of the last 2 months in order to meet his goal if he saves the same amount each month? (Lesson 2.2)

   (A) $150

   (B) $300

   (C) $450

   (D) $600

4. What is the total cost of 0.5 pound of peaches selling for $0.80 per pound and 0.7 pound of oranges selling for $0.90 per pound? (Lesson 4.7)

   (A) $0.51

   (B) $1.02

   (C) $1.03

   (D) $10.30

5. Megan hiked 15.12 miles in 6.3 hours. If Megan hiked the same number of miles each hour, how many miles did she hike each hour? (Lesson 5.6)

   (A) 0.24 miles

   (B) 0.252 miles

   (C) 2.4 miles

   (D) 2.52 miles

6. It is $42\frac{1}{2}$ miles from Eaton to Baxter, and $37\frac{4}{5}$ miles from Baxter to Wellington. How far is it from Eaton to Wellington, if you go by way of Baxter? (Lesson 6.6)

   (A) $4\frac{7}{10}$ miles

   (B) $79\frac{1}{2}$ miles

   (C) $80\frac{3}{10}$ miles

   (D) $80\frac{2}{5}$ miles

Name _____

Fraction Multiplication

Find the product. Write the product in simplest form.

1. $\dfrac{4}{5} \times \dfrac{7}{8} = \dfrac{4 \times 7}{5 \times 8}$

$= \dfrac{28}{40}$

$= \dfrac{7}{10}$

2. $3 \times \dfrac{1}{6}$

3. $\dfrac{5}{9} \times \dfrac{3}{4}$

4. $\dfrac{4}{7} \times \dfrac{1}{2}$

5. $\dfrac{1}{8} \times 20$

6. $\dfrac{4}{5} \times \dfrac{3}{8}$

7. $\dfrac{6}{7} \times \dfrac{7}{9}$

8. $8 \times \dfrac{1}{9}$

9. $\dfrac{1}{14} \times 28$

10. $\dfrac{3}{4} \times \dfrac{1}{3}$

11. Karen raked $\dfrac{3}{5}$ of the yard. Minni raked $\dfrac{1}{3}$ of the amount Karen raked. How much of the yard did Minni rake?

12. In the pet show, $\dfrac{3}{8}$ of the pets are dogs. Of the dogs, $\dfrac{2}{3}$ have long hair. What fraction of the pets are dogs with long hair?

Algebra Evaluate for the given value of the variable.

13. $\dfrac{7}{8} \times c$ for $c = 8$

14. $t \times \dfrac{3}{4}$ for $t = \dfrac{8}{9}$

15. $\dfrac{1}{2} \times s$ for $s = \dfrac{3}{10}$

16. $y \times 6$ for $y = \dfrac{2}{3}$

Problem Solving  REAL WORLD

17. Jason ran $\dfrac{5}{7}$ of the distance around the school track. Sara ran $\dfrac{4}{5}$ of Jason's distance. What fraction of the total distance around the track did Sara run?

18. A group of students attend a math club. Half of the students are boys and $\dfrac{4}{9}$ of the boys have brown eyes. What fraction of the group are boys with brown eyes?

© Houghton Mifflin Harcourt Publishing Company

Lesson Check

1. Fritz attended band practice for $\frac{5}{6}$ hour. Then he went home and practiced for $\frac{2}{5}$ as long as band practice. How many minutes did he practice at home?

   (A) 10 minutes

   (B) 15 minutes

   (C) 20 minutes

   (D) 25 minutes

2. Darlene read $\frac{5}{8}$ of a 56-page book. How many pages did Darlene read?

   (A) 30

   (B) 35

   (C) 40

   (D) 45

Spiral Review

3. What is the quotient of $\frac{18}{1,000}$? (Lesson 5.1)

   (A) 18,000

   (B) 1,800

   (C) 0.18

   (D) 0.018

4. A machine produces 1,000 bowling pins per hour, each valued at $8.37. What is the total value of the pins produced in 1 hour?
   (Lesson 4.1)

   (A) $8.37

   (B) $83.70

   (C) $837.00

   (D) $8,370.00

5. Keith had $8\frac{1}{2}$ cups of flour. He used $5\frac{2}{3}$ cups to make bread. How many cups of flour does Keith have left? (Lesson 6.7)

   (A) $1\frac{5}{6}$ cups

   (B) $2\frac{5}{6}$ cups

   (C) $3\frac{1}{6}$ cups

   (D) $3\frac{1}{3}$ cups

6. The Blue Lake Trail is $11\frac{3}{8}$ miles long. Gemma has hiked $2\frac{1}{2}$ miles each hour for 3 hours. How far is she from the end of the trail? (Lesson 7.3)

   (A) $3\frac{7}{8}$ miles

   (B) $4\frac{1}{2}$ miles

   (C) $4\frac{7}{8}$ miles

   (D) $8\frac{7}{8}$ miles

Name _____

Area and Mixed Numbers

Use the grid to find the area.

1. Let each square represent $\frac{1}{4}$ unit by $\frac{1}{4}$ unit.

   $2\frac{1}{4} \times 1\frac{1}{2} = $ **$3\frac{3}{8}$**

   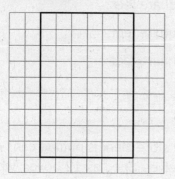

   ___54___ squares cover the diagram.

   Each square is ___$\frac{1}{16}$___ square unit.

   The area of the diagram is

   ___$54 \times \frac{1}{16} = \frac{54}{16} = 3\frac{3}{8}$___ square units.

2. Let each square represent $\frac{1}{3}$ unit by $\frac{1}{3}$ unit.

   $1\frac{2}{3} \times 2\frac{1}{3} = $ _____

   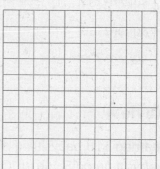

   The area is _____ square units.

Use an area model to solve.

3. $1\frac{1}{8} \times 2\frac{1}{2}$

4. $2\frac{2}{3} \times 1\frac{1}{3}$

5. $1\frac{3}{4} \times 2\frac{1}{2}$

_____   _____   _____

Problem Solving REAL WORLD

6. Ava's bedroom rug is $2\frac{3}{4}$ feet long and $2\frac{1}{2}$ feet wide. What is the area of the rug?

7. A painting is $2\frac{2}{3}$ feet long and $1\frac{1}{2}$ feet high. What is the area of the painting?

_____   _____

Lesson Check

1. The base of a fountain is rectangular. Its dimensions are $1\frac{2}{3}$ feet by $2\frac{2}{3}$ feet. What is the area of the base of the fountain?

   (A) $2\frac{4}{9}$ square feet

   (B) $3\frac{4}{9}$ square feet

   (C) $4\frac{1}{3}$ square feet

   (D) $4\frac{4}{9}$ square feet

2. Bill's living room floor is covered with carpet tiles. Each tile is $1\frac{1}{2}$ feet long by $2\frac{3}{5}$ feet wide. What is the area of one tile?

   (A) $2\frac{3}{10}$ square feet

   (B) $2\frac{9}{10}$ square feet

   (C) $3\frac{9}{10}$ square feet

   (D) $4\frac{1}{10}$ square feet

Spiral Review

3. Lucy earned $18 babysitting on Friday and $20 babysitting on Saturday. On Sunday, she spent half of the money. Which expression matches the words? **(Lesson 1.10)**

   (A) $18 + 20 \div 2$

   (B) $(18 + 20) \div 2$

   (C) $2 \times (18 + 20)$

   (D) $2 \times 18 + 20$

4. A grocery store clerk is putting cans of soup on the shelves. She has 12 boxes, which each contain 24 cans of soup. Altogether, how many cans of soup will the clerk put on the shelves? **(Lesson 1.7)**

   (A) 36

   (B) 208

   (C) 248

   (D) 288

5. Which is the best estimate for the quotient $5,397 \div 62$? **(Lesson 2.5)**

   (A) 80

   (B) 90

   (C) 800

   (D) 900

6. There are 45 vehicles in a parking lot. Three fifths of the vehicles are minivans. How many of the vehicles in the parking lot are minivans? **(Lesson 7.3)**

   (A) 9

   (B) 18

   (C) 27

   (D) 35

Name _____

Compare Mixed Number Factors and Products

Complete the statement with *equal to*, *greater than*, or *less than*.

1. $\frac{2}{3} \times 1\frac{5}{8}$ will be __**less than**__ $1\frac{5}{8}$.

   Think: $1 \times 1\frac{5}{8}$ is $1\frac{5}{8}$.

   Since $\frac{2}{3}$ is less than 1,

   $\frac{2}{3} \times 1\frac{5}{8}$ will be less than $1\frac{5}{8}$.

2. $\frac{5}{5} \times 2\frac{3}{4}$ will be _____ $2\frac{3}{4}$.

3. $3 \times 3\frac{2}{7}$ will be _____ $3\frac{2}{7}$.

4. $9 \times 1\frac{4}{5}$ will be _____ $1\frac{4}{5}$.

5. $1\frac{7}{8} \times 2\frac{3}{8}$ will be _____ $2\frac{3}{8}$.

6. $3\frac{4}{9} \times \frac{5}{9}$ will be _____ $3\frac{4}{9}$.

Problem Solving REAL WORLD

7. Fraser is making a scale drawing of a dog house. The dimensions of the drawing will be $\frac{1}{8}$ of the dimensions of the actual doghouse. The height of the actual doghouse is $36\frac{3}{4}$ inches. Will the dimensions of Fraser's drawing be equal to, greater than, or less than the dimensions of the actual dog house?

8. Jorge has a recipe that calls for $2\frac{1}{3}$ cups of flour. He plans to make $1\frac{1}{2}$ times the recipe. Will the amount of flour Jorge needs be equal to, greater than, or less than the amount of flour his recipe calls for?

_____

_____

Lesson Check

1. Jenna skis $2\frac{1}{4}$ miles in one hour. Her instructor skis $1\frac{1}{3}$ times as far in one hour. Which statement below is true?

   (A) Jenna skis a greater distance than her instructor skis.

   (B) Jenna skis a shorter distance than her instructor skis.

   (C) Jenna skis the same distance as her instructor skis.

   (D) Jenna skis twice the distance her instructor skis.

2. Suppose you multiply a fraction less than 1 by the mixed number $2\frac{3}{4}$. Which statement below will be true?

   (A) The product will be equal to $2\frac{3}{4}$.

   (B) The product will be greater than $2\frac{3}{4}$.

   (C) The product will be less than $2\frac{3}{4}$.

   (D) You cannot make any conclusions about the product.

Spiral Review

3. Rectangular Washington County measures 15.9 miles by 9.1 miles. What is the county's area? **(Lesson 4.7)**

   (A)   1.8 square miles

   (B)   6.8 square miles

   (C)   25 square miles

   (D) 144.69 square miles

4. Marsha jogged 7.8 miles. Erica jogged 0.5 times as far. How far did Erica jog? **(Lesson 4.7)**

   (A)   0.39 mile

   (B)   3.9 miles

   (C)   39 miles

   (D) 390 miles

5. One cookie recipe calls for $2\frac{1}{3}$ cups of sugar. Another cookie recipe calls for $2\frac{1}{2}$ cups of sugar. Tim has 5 cups of sugar. If he bakes both recipes, how much sugar will he have left over? **(Lesson 6.7)**

   (A)   0 cups

   (B) $\frac{1}{6}$ cup

   (C) $1\frac{1}{6}$ cups

   (D) $4\frac{5}{6}$ cups

6. On Monday, it rained $1\frac{1}{4}$ inches. On Tuesday, it rained $\frac{3}{5}$ inch. How much more did it rain on Monday than on Tuesday? **(Lesson 6.7)**

   (A) $\frac{2}{20}$ inch

   (B) $\frac{3}{5}$ inch

   (C) $\frac{13}{20}$ inch

   (D) $1\frac{17}{20}$ inches

Multiply Mixed Numbers

Find the product. Write the product in simplest form.

1. $1\frac{2}{3} \times 4\frac{2}{5}$

$1\frac{2}{3} \times 4\frac{2}{5} = \frac{5}{3} \times \frac{22}{5}$

$\qquad = \frac{110}{15} = \frac{22}{3}$

$\qquad = 7\frac{1}{3}$

2. $1\frac{1}{7} \times 1\frac{3}{4}$

3. $8\frac{1}{3} \times \frac{3}{5}$

4. $2\frac{5}{8} \times 1\frac{2}{3}$

5. $5\frac{1}{2} \times 3\frac{1}{3}$

6. $7\frac{1}{5} \times 2\frac{1}{6}$

7. $\frac{2}{3} \times 4\frac{1}{5}$

8. $2\frac{2}{5} \times 1\frac{1}{4}$

Use the Distributive Property to find the product.

9. $4\frac{2}{5} \times 10$

10. $26 \times 2\frac{1}{2}$

11. $6 \times 3\frac{2}{3}$

Problem Solving

12. Jake can carry $6\frac{1}{4}$ pounds of wood in from the barn. His father can carry $1\frac{5}{7}$ times as much as Jake. How many pounds can Jake's father carry?

13. A glass can hold $3\frac{1}{3}$ cups of water. A bowl can hold $2\frac{3}{5}$ times the amount in the glass. How many cups can a bowl hold?

Lesson Check

1. A vet weighs two puppies. The small puppy weighs $4\frac{1}{2}$ pounds. The large puppy weighs $4\frac{2}{3}$ times as much as the small puppy. How much does the large puppy weigh?

   (A) $16\frac{1}{6}$ pounds

   (B) 19 pounds

   (C) 21 pounds

   (D) 25 pounds

2. Becky lives $5\frac{5}{8}$ miles from school. Steve lives $1\frac{5}{9}$ times as far from school as Becky. How far does Steve live from school?

   (A) 12 miles

   (B) $8\frac{3}{4}$ miles

   (C) 6 miles

   (D) $5\frac{3}{16}$ miles

Spiral Review

3. Craig scored 12 points in a game. Marla scored twice as many points as Craig but 5 fewer points than Nelson scored. How many points did Nelson score? **(Lesson 1.10)**

   (A) $2 \times 12 + 5$

   (B) $2 \times 12 - 5$

   (C) $\frac{1}{2} \times 12 + 5$

   (D) $2 \times (12 + 5)$

4. Yvette earned $66.00 for 8 hours of work. Lizbeth earned $68.80 working the same amount of time. How much more per hour did Lizbeth earn than Yvette earned?

   (Lesson 5.4)

   (A) $0.35

   (B) $0.45

   (C) $2.80

   (D) $8.25

5. What is the least common denominator of the four fractions listed below? **(Lesson 6.4)**

   $$20\frac{7}{10} \qquad 20\frac{3}{4} \qquad 18\frac{9}{10} \qquad 20\frac{18}{25}$$

   (A) 2

   (B) 40

   (C) 50

   (D) 100

6. Three girls searched for geodes in the desert. Corinne collected $11\frac{1}{8}$ pounds, Ellen collected $4\frac{5}{8}$ pounds, and Leonda collected $3\frac{3}{4}$ pounds. How much more did Corinne collect than the other two girls combined?

   (Lesson 6.6)

   (A) $2\frac{1}{2}$ pounds

   (B) $2\frac{3}{4}$ pounds

   (C) $2\frac{7}{8}$ pounds

   (D) $3\frac{3}{4}$ pounds

Name _____

Problem Solving • Find Unknown Lengths

1. Kamal's bedroom has an area of 120 square feet. The width of the room is $\frac{5}{6}$ the length of the room. What are the dimensions of Kamal's bedroom?

   Guess: $6 \times 20 = 120$
   Check: $\frac{5}{6} \times 20 = 16\frac{2}{3}$; try a longer width.
   Guess: $10 \times 12 = 120$
   Check: $\frac{5}{6} \times 12 = 10$. Correct!

   ## 10 feet by 12 feet

2. Marisol is painting on a piece of canvas that has an area of 180 square inches. The length of the painting is $1\frac{1}{4}$ times the width. What are the dimensions of the painting?

   _____

3. A small plane is flying a banner in the shape of a rectangle. The area of the banner is 144 square feet. The width of the banner is $\frac{1}{4}$ the length of the banner. What are the dimensions of the banner?

   _____

4. An artificial lake is in the shape of a rectangle and has an area of $\frac{9}{20}$ square mile. The width of the lake is $\frac{1}{5}$ the length of the lake. What are the dimensions of the lake?

   _____

Lesson Check

1. Consuelo's living room is in the shape of a rectangle and has an area of 360 square feet. The width of the living room is $\frac{5}{8}$ its length. What is the length of the living room?

   (A) 15 feet

   (B) 18 feet

   (C) 20 feet

   (D) 24 feet

2. A rectangular park has an area of $\frac{2}{3}$ square mile. The length of the park is $2\frac{2}{3}$ the width of the park. What is the width of the park?

   (A) $\frac{1}{2}$ mile

   (B) $\frac{2}{3}$ mile

   (C) $1\frac{1}{3}$ miles

   (D) 2 miles

Spiral Review

3. Debra babysits for $3\frac{1}{2}$ hours on Friday and $1\frac{1}{2}$ times as long on Saturday. Which statement below is true? **(Lesson 7.8)**

   (A) Debra babysat more hours on Friday than on Saturday.

   (B) Debra babysat the same number of hours on Friday and Saturday.

   (C) Debra babysat 3 times as many hours on Friday than on Saturday.

   (D) Debra babysat more hours on Saturday than on Friday.

4. Tory practiced her basketball shots for $\frac{2}{3}$ hour. Tim practiced his basketball shots $\frac{3}{4}$ as much time as Tory did. How long did Tim practice his basketball shots? **(Lesson 7.6)**

   (A) $\frac{1}{2}$ hour

   (B) $\frac{1}{3}$ hour

   (C) $\frac{1}{4}$ hour

   (D) $\frac{1}{6}$ hour

5. Leah bought $4\frac{1}{2}$ pounds of grapes. Of the grapes she bought, $1\frac{7}{8}$ pounds were red grapes. The rest were green grapes. How many pounds of green grapes did Leah buy? **(Lesson 6.7)**

   (A) $2\frac{3}{8}$ pounds

   (B) $2\frac{5}{8}$ pounds

   (C) $3\frac{3}{8}$ pounds

   (D) $3\frac{5}{8}$ pounds

6. To which place value is the following number rounded? **(Lesson 3.4)**

   $$5.927 \text{ to } 5.93$$

   (A) ones

   (B) tenths

   (C) hundredths

   (D) thousandths

© Houghton Mifflin Harcourt Publishing Company

Name _____

Chapter 7 Extra Practice

Lesson 7.1

Use a model to solve.

1. $\frac{2}{5} \times 10 =$ _____

2. $\frac{1}{4} \times 24 =$ _____

3. $\frac{3}{7} \times 28 =$ _____

4. $\frac{4}{9} \times 18 =$ _____

5. $\frac{2}{3} \times 21 =$ _____

6. $\frac{4}{11} \times 22 =$ _____

Lessons 7.2 - 7.4, 7.6

Find the product. Write the product in simplest form.

1. $\frac{3}{7} \times 9 =$ _____

2. $8 \times \frac{1}{5} =$ _____

3. $\frac{4}{9} \times 11 =$ _____

4. $2 \times \frac{2}{5} =$ _____

5. $\frac{3}{4} \times 5 =$ _____

6. $3 \times \frac{6}{8} =$ _____

7. $\frac{1}{3} \times \frac{4}{5} =$ _____

8. $\frac{2}{7} \times \frac{3}{8} =$ _____

9. $\frac{4}{9} \times \frac{1}{3} =$ _____

10. $3 \times \frac{1}{9} =$ _____

11. $\frac{5}{7} \times \frac{5}{9} =$ _____

12. $\frac{1}{8} \times \frac{2}{4} =$ _____

13. At the aquarium, $\frac{3}{4}$ of the animals are fish. Of the fish, $\frac{1}{3}$ are clown fish. What fraction of the animals at the aquarium are clown fish?

14. Four hamburgers each contain $\frac{1}{5}$ pound of beef. Altogether, how much beef do the hamburgers contain?

© Houghton Mifflin Harcourt Publishing Company

Lessons 7.5, 7.8

Complete the statement with *equal to*, *greater than*, or *less than*.

1. $\frac{3}{7} \times \frac{4}{9}$ will be _____ than $\frac{4}{9}$.

2. $6 \times \frac{5}{7}$ will be _____ than $\frac{5}{7}$.

3. $\frac{3}{3} \times 5\frac{1}{9}$ will be _____ than $5\frac{1}{9}$.

4. $7 \times 2\frac{5}{9}$ will be _____ than $2\frac{5}{9}$.

Lesson 7.9

Find the product. Write the product in simplest form.

1. $\frac{1}{4} \times 2\frac{1}{2}$

2. $4\frac{1}{2} \times 1\frac{2}{3}$

3. $2\frac{1}{2} \times 1\frac{1}{5}$

4. $3\frac{2}{5} \times 1\frac{2}{3}$

_____ _____ _____ _____

5. $2\frac{3}{5} \times 3\frac{1}{8}$

6. $5 \times 3\frac{1}{3}$

7. $2\frac{3}{5} \times 9\frac{1}{2}$

8. $1\frac{1}{4} \times 8\frac{2}{3}$

_____ _____ _____ _____

Use the Distributive Property to find the product.

9. $15 \times 3\frac{1}{5}$ _____

10. $2\frac{3}{7} \times 21$ _____

Lesson 7.10

1. Gabriella wants to tile a room with an area of 320 square feet. The width of the room is $\frac{4}{5}$ its length. What are the length and width of the room?

2. Akio wants to make a scale drawing that is $\frac{1}{5}$ the size of an original painting. A ship in the original painting is 14 inches long. How long will the ship in Akio's drawing be?

_____ _____

School-Home Letter

Dear Family,

Throughout the next few weeks, our math class will be learning about dividing whole numbers by unit fractions and dividing unit fractions by whole numbers. We will also learn how a fraction represents division.

You can expect to see homework with real-world problems that involve division with fractions.

Here is a sample of how your child will be taught to use a model to divide by a fraction.

Vocabulary

dividend The number that is to be divided in a division problem

equation An algebraic or numerical sentence that shows that two quantities are equal

fraction A number that names a part of a whole or a part of a group

🔒 MODEL   Draw a Diagram to Divide

Sue makes 3 waffles. She divides each waffle into fourths. How many $\frac{1}{4}$-waffle pieces does she have?

Divide. $3 \div \frac{1}{4}$

STEP 1
Draw 3 circles to represent the waffles. Draw lines to divide each circle into fourths.

STEP 2
To find $3 \div \frac{1}{4}$, multiply 3 by the number of fourths in each circle.

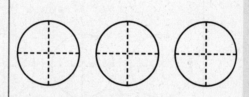

$3 \times 4 = 12$

So, Sue has 12 one-fourth-waffle pieces.

Tips

Using Multiplication to Check

You can use multiplication to check the answer to a division problem involving fractions.

To check the answer in the sample, multiply $\frac{1}{4}$ by 12 and compare the product to the dividend, 3.

$12 \times \frac{1}{4} = \frac{12}{4}$, or 3

Activity

Use real-world division situations such as sharing a pizza, pie, or orange equally to help your child practice division with fractions.

Carta para la casa

Vocabulario

dividendo El número que se va a dividir en un problema de división

ecuación Una expresión algebraica o numérica que muestra que dos cantidades son iguales

fracción Un número que nombra parte de un todo o parte de un grupo

Querida familia,

Durante las próximas semanas, en la clase de matemáticas aprenderemos a dividir números enteros entre fracciones unitarias y fracciones unitarias entre números enteros. También aprenderemos de qué manera una fracción representa una división.

Llevaré a casa tareas para resolver problemas de la vida diaria que incluyan la división con fracciones.

Este es un ejemplo de cómo aprenderemos a usar un modelo para dividir entre una fracción.

🔑 MODELO Dibujar un diagrama para dividir

Sue hace 3 wafles y los divide en porciones de 1/4.
¿Cuántas porciones tiene?

Pistas

Usar la multiplicación para comprobar

Puedes usar la multiplicación para comprobar la respuesta de un problema de división con fracciones.

Para comprobar la respuesta del ejemplo, multiplica $\frac{1}{4}$ por 12 y compara el producto con el dividendo, 3.

$12 \times \frac{1}{4} = \frac{12}{4}$, o 3

PASO 1

Dibuja 3 círculos para representar los wafles. Dibuja líneas que dividan cada círculo en cuartos.

PASO 2

Para hallar $3 \div \frac{1}{4}$, multiplica 3 por el número de cuartos en cada círculo.

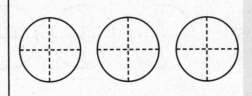

$3 \times 4 = 12$

Por tanto, Sue puede hacer 12 porciones de $\frac{1}{4}$.

Actividad

Use situaciones de la vida diaria que involucren divisiones, como compartir una pizza, un pay o una naranja en partes iguales para ayudar a su hijo o hija a practicar la división con fracciones.

Name _____

Divide Fractions and Whole Numbers

Divide and check the quotient.

1.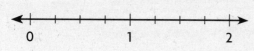

$2 \div \frac{1}{3} = \underline{\quad 6 \quad}$ because $\underline{\quad 6 \quad} \times \frac{1}{3} = 2$.

2.

$2 \div \frac{1}{4} = \underline{\qquad}$ because $\underline{\qquad} \times \frac{1}{4} = 2$.

3.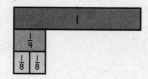

$\frac{1}{4} \div 2 = \underline{\qquad}$ because $\underline{\qquad} \times 2 = \frac{1}{4}$.

Divide. Draw a number line or use fraction strips.

4. $1 \div \frac{1}{5} = \underline{\qquad}$

5. $\frac{1}{6} \div 3 = \underline{\qquad}$

6. $4 \div \frac{1}{6} = \underline{\qquad}$

7. $3 \div \frac{1}{3} = \underline{\qquad}$

8. $\frac{1}{4} \div 6 = \underline{\qquad}$

9. $5 \div \frac{1}{4} = \underline{\qquad}$

Problem Solving REAL WORLD

10. Amy can run $\frac{1}{10}$ mile per minute. How many minutes will it take Amy to run 3 miles?

11. Jeremy has 3 yards of ribbon to use for wrapping gifts. He cuts the ribbon into pieces that are $\frac{1}{4}$ yard long. How many pieces of ribbon does Jeremy have?

_____

_____

Lesson Check

1. Kaley cuts half of a loaf of bread into 4 equal parts. What fraction of the whole loaf does each of the 4 parts represent?

   Ⓐ $\frac{1}{8}$

   Ⓑ $\frac{1}{6}$

   Ⓒ $\frac{1}{4}$

   Ⓓ $\frac{1}{2}$

2. When you divide a fraction less than 1 by a whole number greater than 1, how does the quotient compare to the dividend?

   Ⓐ The quotient is greater than the dividend.

   Ⓑ The quotient is less than the dividend.

   Ⓒ The quotient is equal to the dividend.

   Ⓓ There is not enough information to answer the question.

Spiral Review

3. A recipe for chicken and rice calls for $3\frac{1}{2}$ pounds of chicken. Lisa wants to adjust the recipe so that it yields $1\frac{1}{2}$ times as much chicken and rice. How much chicken will she need? **(Lesson 7.9)**

   Ⓐ 2 pounds

   Ⓑ $2\frac{1}{3}$ pounds

   Ⓒ 5 pounds

   Ⓓ $5\frac{1}{4}$ pounds

4. Tim and Sue share a pizza. Tim eats $\frac{2}{3}$ of the pizza. Sue eats half as much of the pizza as Tim does. What fraction of the pizza does Sue eat? **(Lesson 7.6)**

   Ⓐ $\frac{1}{3}$

   Ⓑ $\frac{1}{2}$

   Ⓒ $\frac{3}{5}$

   Ⓓ $\frac{2}{3}$

5. In gym class, you run $\frac{3}{5}$ mile. Your coach runs 10 times that distance each day. How far does your coach run each day? **(Lesson 7.3)**

   Ⓐ $\frac{7}{5}$ miles

   Ⓑ $2\frac{3}{5}$ miles

   Ⓒ 3 miles

   Ⓓ 6 miles

6. Sterling plants a tree that is $4\frac{3}{4}$ feet tall. One year later, the tree is $5\frac{2}{5}$ feet tall. How many feet did the tree grow? **(Lesson 6.7)**

   Ⓐ $\frac{13}{20}$ foot

   Ⓑ 8 feet

   Ⓒ $10\frac{3}{20}$ feet

   Ⓓ 13 feet

Name _____

Problem Solving • Use Multiplication

1. Sebastian bakes 4 pies and cuts each pie into sixths. How many $\frac{1}{6}$-pie slices does he have?

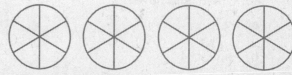

To find the total number of sixths in the 4 pies, multiply 4 by the number of sixths in each pie.

$$4 \div \frac{1}{6} = 4 \times 6 = 24 \text{ one-sixth-pie slices}$$

2. Ali has 2 vegetable pizzas that she cuts into eighths. How many $\frac{1}{8}$-size pieces does she have?

_____

3. A baker has 6 loaves of bread. Each loaf weighs 1 pound. He cuts each loaf into thirds. How many $\frac{1}{3}$-pound loaves of bread does the chef now have?

_____

4. Suppose the baker has 4 loaves of bread and cuts the loaves into halves. How many $\frac{1}{2}$-pound loaves of bread would the baker have?

_____

5. Madalyn has 3 watermelons that she cuts into halves to give to her neighbors. How many neighbors will get a $\frac{1}{2}$-size piece of watermelon?

_____

6. A landscaper had 5 tons of rock to build decorative walls. He used $\frac{1}{4}$ ton of rock for each wall. How many decorative walls did he build?

_____

Lesson Check

1. Julia has 12 pieces of fabric and cuts each piece into fourths. How many $\frac{1}{4}$ pieces of fabric does she have?

   (A) 3
   (B) 4
   (C) 24
   (D) 48

2. Josue has 3 cheesecakes that he cuts into thirds. How many $\frac{1}{3}$-size cheesecake pieces does he have?

   (A) 9
   (B) 6
   (C) 3
   (D) 1

Spiral Review

3. Which of the following multiplication sentences can you use to help you find the quotient $6 \div \frac{1}{4}$? **(Lesson 8.1)**

   (A) $6 \times \frac{1}{4} = \frac{6}{4}$
   (B) $\frac{1}{6} \times 4 = \frac{4}{6}$
   (C) $\frac{1}{6} \times \frac{1}{4} = \frac{1}{24}$
   (D) $24 \times \frac{1}{4} = 6$

4. Ellie uses 12.5 pounds of potatoes to make mashed potatoes. She uses one-tenth as many pounds of butter as potatoes. How many pounds of butter does Ellie use?
   (Lesson 5.1)

   (A)   0.125 pound
   (B)   1.25 pounds
   (C)   125 pounds
   (D) 1,250 pounds

5. Tiffany collects perfume bottles. She has 99 bottles in her collection. Two-thirds of her perfume bottles are made of crystal. How many of the perfume bottles in Tiffany's collection are made of crystal? **(Lesson 7.1)**

   (A) 11
   (B) 33
   (C) 66
   (D) 99

6. Stephen makes a blueberry pie and cuts it into 6 slices. He eats $\frac{1}{3}$ of the pie over the weekend. How many slices of pie does Stephen eat over the weekend? **(Lesson 7.3)**

   (A) 6
   (B) 3
   (C) 2
   (D) 1

Connect Fractions to Division

Complete the number sentence to solve.

1. Six students share 8 apples equally. How many apples does each student get?

$$8 \div 6 = \underline{\dfrac{8}{6}, \text{ or } 1\dfrac{1}{3}}$$

2. Ten boys share 7 cereal bars equally. What fraction of a cereal bar does each boy get?

$7 \div 10 = \underline{\hspace{3cm}}$

3. Eight friends share 12 pies equally. How many pies does each friend get?

$12 \div 8 = \underline{\hspace{3cm}}$

4. Three girls share 8 yards of fabric equally. How many yards of fabric does each girl get?

$8 \div 3 = \underline{\hspace{3cm}}$

5. Five bakers share 2 loaves of bread equally. What fraction of a loaf of bread does each baker get?

$2 \div 5 = \underline{\hspace{3cm}}$

6. Nine friends share 6 cookies equally. What fraction of a cookie does each friend get?

$6 \div 9 = \underline{\hspace{3cm}}$

7. Twelve students share 3 pizzas equally. What fraction of a pizza does each student get?

$3 \div 12 = \underline{\hspace{3cm}}$

8. Three sisters share 5 sandwiches equally. How many sandwiches does each sister get?

$5 \div 3 = \underline{\hspace{3cm}}$

Problem Solving REAL WORLD

9. There are 12 students in a jewelry-making class and 8 sets of charms. What fraction of a set of charms will each student get?

_____

_____

10. Five friends share 6 cheesecakes equally. How many cheesecakes will each friend get?

_____

_____

Lesson Check

1. Eight friends share 4 bunches of grapes equally. What fraction of a bunch of grapes does each friend get?

   (A) $\frac{1}{8}$

   (B) $\frac{1}{4}$

   (C) $\frac{1}{2}$

   (D) 2

2. Ten students share 8 pieces of poster board equally. What fraction of a piece of poster board does each student get?

   (A) $1\frac{4}{5}$

   (B) $1\frac{1}{4}$

   (C) $\frac{4}{5}$

   (D) $\frac{5}{9}$

Spiral Review

3. Arturo has a log that is 4 yards long. He cuts the log into pieces that are $\frac{1}{3}$-yard long. How many pieces will Arturo have? **(Lesson 8.1)**

   (A) $\frac{3}{4}$

   (B) $\frac{4}{3}$

   (C) 6

   (D) 12

4. Vu has 2 pizzas that he cuts into sixths. How many $\frac{1}{6}$-size pieces does he have? **(Lesson 8.2)**

   (A) 12

   (B) 6

   (C) 3

   (D) $\frac{1}{3}$

5. Kayaks rent for $35 per day. Which expression can you use to find the cost in dollars of renting 3 kayaks for a day? **(Lesson 1.3)**

   (A) $(3 + 30) + (3 + 5)$

   (B) $(3 \times 30) + (3 \times 5)$

   (C) $(3 + 30) \times (3 + 5)$

   (D) $(3 \times 30) \times (3 \times 5)$

6. Louisa is 152.7 centimeters tall. Her younger sister is 8.42 centimeters shorter than she is. How tall is Louisa's younger sister? **(Lesson 3.9)**

   (A) 6.85 cm

   (B) 144.28 cm

   (C) 144.38 cm

   (D) 154.28 cm

Name _____

Fraction and Whole-Number Division

Write a related multiplication sentence to solve.

1. $3 \div \frac{1}{2}$

2. $\frac{1}{5} \div 3$

3. $2 \div \frac{1}{8}$

4. $\frac{1}{3} \div 4$

$\underline{3 \times 2 = 6}$

_____

_____

_____

5. $5 \div \frac{1}{4}$

6. $\frac{1}{2} \div 2$

7. $\frac{1}{4} \div 6$

8. $6 \div \frac{1}{5}$

_____

_____

_____

_____

9. $\frac{1}{5} \div 5$

10. $4 \div \frac{1}{8}$

11. $\frac{1}{3} \div 7$

12. $9 \div \frac{1}{2}$

_____

_____

_____

_____

Problem Solving

13. Isaac has a piece of rope that is 5 yards long. Into how many $\frac{1}{2}$-yard pieces of rope can Isaac cut the rope?

14. Two friends share $\frac{1}{2}$ of a pineapple equally. What fraction of a whole pineapple does each friend get?

_____

_____

Lesson Check

1. Sean divides 8 cups of granola into $\frac{1}{4}$-cup servings. How many servings of granola does he have?

   (A) 32

   (B) 16

   (C) 2

   (D) $\frac{1}{2}$

2. Brandy solved $\frac{1}{6} \div 5$ by using a related multiplication expression. Which multiplication expression did she use?

   (A) 6×5

   (B) $6 \times \frac{1}{5}$

   (C) $\frac{1}{6} \times 5$

   (D) $\frac{1}{6} \times \frac{1}{5}$

Spiral Review

3. Nine friends share 12 pounds of pecans equally. How many pounds of pecans does each friend get? **(Lesson 8.3)**

   (A) $\frac{3}{4}$ pound

   (B) $1\frac{1}{3}$ pounds

   (C) $1\frac{1}{2}$ pounds

   (D) $1\frac{2}{3}$ pounds

4. A scientist has $\frac{2}{3}$ liter of solution. He uses $\frac{1}{2}$ of the solution for an experiment. How much solution does the scientist use for the experiment? **(Lesson 7.6)**

   (A) $\frac{1}{6}$ liter

   (B) $\frac{1}{4}$ liter

   (C) $\frac{1}{3}$ liter

   (D) $\frac{1}{2}$ liter

5. Naomi needs 2 cups of sugar for a cake she is baking. She only has a $\frac{1}{4}$-cup measuring cup. How many times will Naomi need to fill the measuring cup to get 2 cups of sugar?

   (Lesson 8.2)

   (A) 2

   (B) 4

   (C) 6

   (D) 8

6. Michaela caught 3 fish, which weigh a total of $19\frac{1}{2}$ pounds. One fish weighs $7\frac{5}{8}$ pounds and another weighs $5\frac{3}{4}$ pounds. How much does the third fish weigh? **(Lesson 6.9)**

   (A) $6\frac{1}{8}$ pounds

   (B) $6\frac{5}{8}$ pounds

   (C) $7\frac{1}{8}$ pounds

   (D) $7\frac{5}{8}$ pounds

© Houghton Mifflin Harcourt Publishing Company

Interpret Division with Fractions

Write an equation to represent the problem. Then solve.

1. Daniel has a piece of wire that is $\frac{1}{2}$ yard long. He cuts the wire into 3 equal pieces. What fraction of a yard is each piece?

$$\frac{1}{2} \div 3 = n; \frac{1}{2} \times \frac{1}{3} = n;$$

$$n = \frac{1}{6}; \frac{1}{6} \text{ yard}$$

2. Vita has a piece of ribbon that is 5 meters long. She cuts the ribbon into pieces that are each $\frac{1}{3}$ meter long. How many pieces does she cut?

_____

Draw a diagram to represent the problem. Then solve.

3. Leah has 3 muffins. She cuts each muffin into fourths. How many $\frac{1}{4}$-muffin pieces does she have?

4. Two friends share $\frac{1}{4}$ gallon of lemonade equally. What fraction of the gallon of lemonade does each friend get?

_____   _____

5. Write a story problem to represent $3 \div \frac{1}{2}$.

_____

_____

6. Write a story problem to represent $\frac{1}{4} \div 2$.

_____

_____

Problem Solving REAL WORLD

7. Spencer has $\frac{1}{3}$ pound of nuts. He divides the nuts equally into 4 bags. What fraction of a pound of nuts is in each bag?

8. Humma has 3 apples. She slices each apple into eighths. How many $\frac{1}{8}$-apple slices does she have?

_____   _____

Lesson Check

1. Abigail has $\frac{1}{2}$ gallon of orange juice. She pours the same amount of the juice into each of 6 glasses. Which equation represents the fraction of a gallon of orange juice in each glass?

   (A) $6 \div \frac{1}{2} = n$

   (B) $6 \div 2 = n$

   (C) $\frac{1}{2} \div \frac{1}{6} = n$

   (D) $\frac{1}{2} \div 6 = n$

2. Which situation can be represented by $4 \div \frac{1}{2}$?

   (A) Riley has a piece of wire that is $\frac{1}{2}$ yard long. He cuts it into fourths. How long is each piece of wire?

   (B) Riley has a piece of wire that is 4 yards long. He cuts it into pieces that are $\frac{1}{2}$ yard long. How many pieces of wire does Riley have?

   (C) Riley has 4 pieces of wire. Each piece is $\frac{1}{2}$ yard long. How much wire does Riley have in all?

   (D) Riley has a piece of wire that is 4 yards long. He cuts it in half. How long is each piece of wire?

Spiral Review

3. Hannah buys $\frac{2}{3}$ pound of roast beef. She uses $\frac{1}{4}$ pound to make a sandwich for lunch. How much roast beef does she have left? (Lesson 6.5)

   (A) $\frac{5}{12}$ pound

   (B) $\frac{1}{2}$ pound

   (C) $\frac{11}{12}$ pound

   (D) 2 pounds

4. Alex buys $2\frac{1}{2}$ pounds of grapes. He buys $1\frac{1}{4}$ times as many pounds of apples as grapes. How many pounds of apples does Alex buy? (Lesson 7.9)

   (A) $1\frac{1}{4}$ pounds

   (B) $3\frac{1}{8}$ pounds

   (C) $3\frac{1}{3}$ pounds

   (D) $3\frac{3}{4}$ pounds

5. Maritza's car has 16 gallons of gas in the tank. She uses $\frac{3}{4}$ of the gas. How many gallons of gas does Maritza use? (Lesson 7.3)

   (A) 4 gallons

   (B) $5\frac{1}{4}$ gallons

   (C) 12 gallons

   (D) $21\frac{1}{3}$ gallons

6. Jaime has a board that is 8 feet long. He cuts the board into three equal pieces. How long is each piece? (Lesson 8.3)

   (A) $\frac{3}{8}$ foot

   (B) $1\frac{2}{3}$ feet

   (C) $2\frac{2}{3}$ feet

   (D) 24 feet

Chapter 8 Extra Practice

Lesson 8.1

Divide. Draw a number line or use fraction strips.

1. $2 \div \dfrac{1}{4} =$ _____

2. $\dfrac{1}{7} \div 3 =$ _____

3. $4 \div \dfrac{1}{5} =$ _____

4. $3 \div \dfrac{1}{2} =$ _____

5. $\dfrac{1}{8} \div 5 =$ _____

6. $\dfrac{1}{9} \div 3 =$ _____

7. $5 \div \dfrac{1}{6} =$ _____

8. $8 \div \dfrac{1}{3} =$ _____

9. $\dfrac{1}{5} \div 5 =$ _____

Lesson 8.2

Draw a diagram to solve.

1. A baker has 6 small bags of flour. Each bag weighs 1 pound. She divides each bag into thirds. How many $\frac{1}{3}$-pound bags of flour does the baker have?

2. Merril cuts 6 apple pies into halves. How many $\frac{1}{2}$-size pie pieces does she have?

_____   _____

Lesson 8.3

Complete the number sentence to solve.

1. Three students share 5 peaches equally. How many peaches does each student get?

    $5 \div 3 =$ _____

2. Six friends share 4 sandwiches equally. What fraction of a sandwich does each friend get?

    $4 \div 6 =$ _____

3. Ten cousins share 3 pizzas equally. What fraction of a pizza does each cousin get?

    $3 \div 10 =$ _____

4. Four boys share 9 yards of fishing wire equally. How many yards of fishing wire does each boy get?

    $9 \div 4 =$ _____

Lesson 8.4

Write a related multiplication expression to solve.

1. $6 \div \frac{1}{4}$

2. $9 \div \frac{1}{3}$

3. $\frac{1}{6} \div 7$

4. $\frac{1}{4} \div 10$

_____   _____   _____   _____

Lesson 8.5

1. Write an equation to represent the problem. Then solve.

    Luz has $\frac{1}{3}$ pound of cherries. She divides the cherries equally into 2 bags. What fraction of a pound of cherries is in each bag?

2. Draw a diagram to represent the problem. Then solve.

    Tran has 4 submarine sandwiches. He cuts each sandwich into thirds. How many $\frac{1}{3}$-sandwich pieces does he have?

_____

_____   _____

School-Home Letter

Dear Family,

Throughout the next few weeks, our math class will be working with data and graphs. We will learn how to make and use line plots and line graphs to analyze data and solve problems. We will also learn how to plot and name points on a coordinate grid.

You can expect to see homework that includes making and analyzing line graphs.

Here is a sample of how your child will be taught to interpret line graphs.

Vocabulary

interval The difference between one number and the next on the scale of a graph

line graph A graph that uses line segments to show how data changes over time

scale A series of numbers placed at fixed distances on a graph to help label the graph

x-axis The horizontal number line on a coordinate plane

x-coordinate The first number in an ordered pair, which tells the distance to move right or left from (0, 0)

y-axis The vertical number line on a coordinate plane

y-coordinate The second number in an ordered pair, which tells the distance to move up or down from (0, 0)

🔒 MODEL Analyze Line Graphs

This is how we will analyze line graphs.

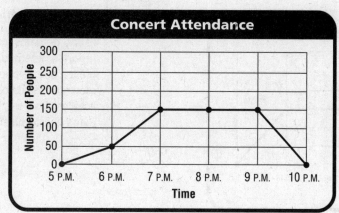

Tips

Choose an Appropriate Graph

The type of data being reported will help determine what type of graph can be used to visually display the data. Line graphs are a good way to display data that change over time.

Use the graph to identify between what times the concert attendance increased the most.

STEP 1 Look at each segment in the graph.

STEP 2 Find the segment that shows the greatest increase in number of people between two consecutive points.

The greatest increase in the number of people occurred between 6 P.M. and 7 P.M.

Activity

Look through a few newspapers or magazines to find data displays. Then work together to write and answer questions about the information displayed.

Carta para la casa

© Houghton Mifflin Harcourt Publishing Company

Vocabulario

intervalo La diferencia entre un número y el siguiente en la escala de una gráfica

gráfica lineal Una gráfica que usa segmentos para mostrar cómo los datos cambian con el tiempo

escala Una serie de números colocados a distancias fijas en una gráfica que permite rotular la gráfica

eje de la x La recta numérica horizontal en un plano de coordenadas

eje de la y La recta numérica vertical en un plano de coordenadas

coordenada x El primer número de un par ordenado que muestra la distancia para moverse hacia la derecha o la izquierda desde (0, 0)

coordenada y El segundo número de un par ordenado que muestra la distancia para moverse hacia arriba o hacia abajo desde (0, 0)

Querida familia,

Durante las próximas semanas, en la clase de matemáticas vamos a trabajar con datos y gráficas. Aprenderemos cómo hacer y usar diagramas de puntos y gráficas lineales para analizar datos y resolver problemas. También aprenderemos a anotar y nombrar los puntos en una cuadrícula de coordenadas.

Llevaré a la casa tareas para aprender a hacer y analizar gráficas lineales.

Este es un ejemplo de la manera como aprenderemos a interpretar gráficas lineales.

🔑 MODELO   Interpretar gráficas lineales

Así es como interpretamos las gráficas lineales.

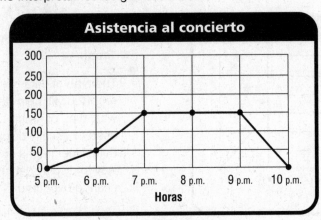

Asistencia al concierto

Pistas

Elegir una gráfica adecuada

El tipo de datos que se van a reportar, determina el tipo de gráfica que se puede usar para mostrar esos datos. Las gráficas lineales nos permiten mostrar datos que cambian con el tiempo.

Usa la gráfica para identificar entre qué horas aumentó más la asistencia al concierto.

PASO 1 Analiza cada segmento de la gráfica.

PASO 2 Halla el segmento que muestre el mayor aumento de la cantidad de personas entre dos puntos consecutivos.
El mayor aumento en la cantidad de personas ocurrió entre las 6 y las 7 P.M.

Actividad

Busquen datos presentados de diferentes formas en periódicos y revistas. Después escriban y contesten juntos preguntas sobre la información que se muestra.

Name _____

Line Plots

Use the data to complete the line plot. Then answer the questions.

A clerk in a health food store makes bags of trail mix. The amount of trail mix in each bag is listed below.

$\frac{1}{4}$ lb, $\frac{1}{4}$ lb, $\frac{3}{4}$ lb, $\frac{1}{2}$ lb, $\frac{1}{4}$ lb, $\frac{3}{4}$ lb,

$\frac{3}{4}$ lb, $\frac{3}{4}$ lb, $\frac{1}{2}$ lb, $\frac{1}{4}$ lb, $\frac{1}{2}$ lb, $\frac{1}{2}$ lb

1. What is the combined weight of the $\frac{1}{4}$-lb bags? __**1 lb**__

   Think: There are four $\frac{1}{4}$-pound bags.

2. What is the combined weight of the $\frac{1}{2}$-lb bags? _____

3. What is the combined weight of the $\frac{3}{4}$-lb bags? _____

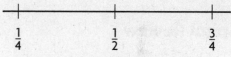

Weight of Trail Mix (in pounds)

4. What is the total weight of the trail mix used in

   all the bags? _____

5. What is the average amount of trail mix in each bag? _____

Julie uses crystals to make a bracelet. The lengths of the crystals are shown below.

$\frac{1}{2}$ in., $\frac{5}{8}$ in., $\frac{3}{4}$ in., $\frac{1}{2}$ in., $\frac{3}{8}$ in., $\frac{1}{2}$ in., $\frac{3}{4}$ in.,

$\frac{3}{8}$ in., $\frac{3}{4}$ in., $\frac{5}{8}$ in., $\frac{1}{2}$ in., $\frac{3}{8}$ in., $\frac{5}{8}$ in., $\frac{3}{4}$ in.

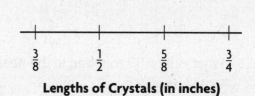

Lengths of Crystals (in inches)

6. What is the combined length of the $\frac{1}{2}$-in. crystals? _____

7. What is the combined length of the $\frac{5}{8}$-in. crystals? _____

8. What is the total length of all the crystals in the bracelet? _____

9. What is the average length of each crystal in the bracelet? _____

Lesson Check

A baker uses different amounts of salt when she bakes loaves of bread, depending on which recipe she is following. The amount of salt called for in each recipe is shown on the line plot.

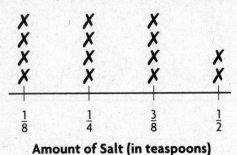

Amount of Salt (in teaspoons)

1. Based on the line plot, how many recipes call for more than $\frac{1}{4}$ tsp of salt?

   (A) 4   (C) 8

   (B) 6   (D) 12

2. What is the average amount of salt called for in each recipe?

   (A) $\frac{1}{8}$ tsp   (C) $\frac{2}{7}$ tsp

   (B) $\frac{1}{4}$ tsp   (D) $\frac{1}{2}$ tsp

Spiral Review

3. Ramona had $8\frac{3}{8}$ in. of ribbon. She used $2\frac{1}{2}$ in. for an art project. How many inches of ribbon does she have left? Find the difference in simplest form. **(Lesson 6.7)**

   (A) $5\frac{1}{8}$ in.

   (B) $5\frac{7}{8}$ in.

   (C) $6\frac{1}{8}$ in.

   (D) $6\frac{1}{6}$ in.

4. Ben bought $\frac{1}{2}$ pound of cheese for 3 sandwiches. If he puts the same amount of cheese on each sandwich, how much cheese will each sandwich have? **(Lesson 8.4)**

   (A) $\frac{1}{6}$ lb

   (B) $\frac{2}{3}$ lb

   (C) $1\frac{1}{2}$ lb

   (D) 6 lb

5. What is 92.583 rounded to the nearest tenth? **(Lesson 3.4)**

   (A) 90

   (B) 92.5

   (C) 92.58

   (D) 92.6

6. In Yoshi's garden, $\frac{3}{4}$ of the flowers are tulips. Of the tulips, $\frac{2}{3}$ are yellow. What fraction of the flowers in Yoshi's garden are yellow tulips? **(Lesson 7.6)**

   (A) $\frac{1}{12}$

   (B) $\frac{5}{12}$

   (C) $\frac{1}{2}$

   (D) $\frac{5}{7}$

Ordered Pairs

Use Coordinate Grid A to write an ordered pair
for the given point.

1. A **(2, 3)**   2. B

3. C   4. D

5. E   6. F

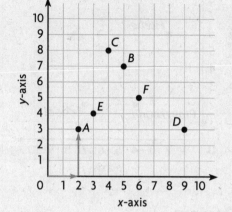

Coordinate Grid A

Plot and label the points on Coordinate Grid B.

7. N (7, 3)   8. R (0, 4)

9. O (8, 7)   10. M (2, 1)

11. P (5, 6)   12. Q (1, 5)

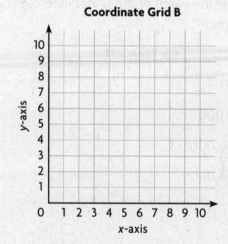

Coordinate Grid B

Problem Solving REAL WORLD

Use the map for 13–14.

13. Which building is located at (5, 6)?

_____

14. What is the distance between Kip's Pizza and
the bank?

_____

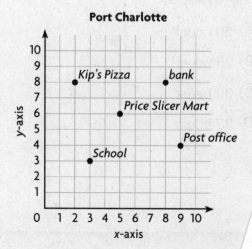

Port Charlotte

Lesson Check

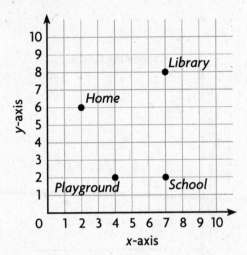

1. Which ordered pair describes the location of the playground?

    (A) (2, 4) (C) (3, 1)

    (B) (4, 2) (D) (1, 3)

2. What is the distance between the school and the library?

    (A) 5 units (C) 7 units

    (B) 6 units (D) 9 units

Spiral Review

3. What is the value of the underlined digit?
   (Lesson 1.2)

   45,7<u>6</u>9,331

   (A) 60

   (B) 6,000

   (C) 60,000

   (D) 70,000

4. Andrew charges $18 for each lawn he mows. Suppose he mows 17 lawns per month. How much money will Andrew make per month?
   (Lesson 1.7)

   (A) $305

   (B) $306

   (C) $350

   (D) $360

5. Harlow can bicycle at a rate of 18 miles per hour. How many hours would it take him to bicycle a stretch of road that is 450 miles long? (Lesson 2.6)

   (A) 20 hours

   (B) 25 hours

   (C) 30 hours

   (D) 35 hours

6. Molly uses 192 beads to make a bracelet and a necklace. It takes 5 times as many beads to make a necklace than it does to make a bracelet. How many beads are used to make the necklace? (Lesson 2.9)

   (A) 32

   (B) 37

   (C) 160

   (D) 165

Graph Data

Graph the data on the coordinate grid.

1.

Outdoor Temperature					
Hour	1	3	5	7	9
Temperature (°F)	61	65	71	75	77

a. Write the ordered pairs for each point.

_____

_____

b. How would the ordered pairs be different if the outdoor temperature were recorded every hour for 4 consecutive hours?

_____

_____

_____

_____

Outdoor Temperature

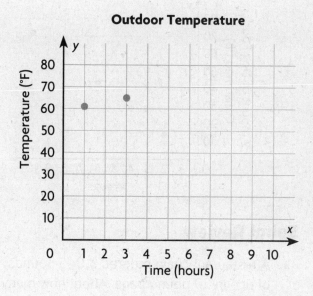

Problem Solving

2.

Windows Repaired					
Day	1	2	3	4	5
Total Number Repaired	14	30	45	63	79

a. Write the ordered pairs for each point.

_____

_____

b. What does the ordered pair (2, 30) tell you about the number of windows repaired?

_____

_____

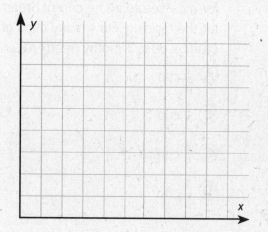

Lesson Check

Amount of Dog Food Consumed

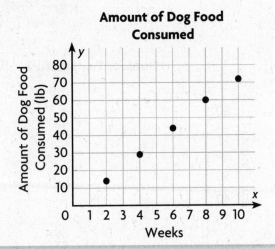

1. About how many weeks did it take for the dog to consume 45 pounds of food?

   (A) 4 weeks   (C) 6 weeks

   (B) 5 weeks   (D) 7 weeks

2. By the end of Week 8, how much food had the dog consumed?

   (A) 29 pounds   (C) 60 pounds

   (B) 44 pounds   (D) 72 pounds

Spiral Review

3. A restaurant chain ordered 3,945 pounds of rice in 20-pound bags. About how many 20-pound bags of rice did the chain order? (Lesson 2.5)

   (A)   4,000

   (B)   2,000

   (C)   200

   (D)   20

4. The population of Linton is 12 times as great as the population of Ellmore. The combined population of both towns is 9,646 people. What is the population of Linton? (Lesson 2.9)

   (A)   742

   (B)   804

   (C) 8,904

   (D) 9,634

5. Timothy needs $\frac{1}{2}$ cup of bread crumbs for a casserole and $\frac{1}{3}$ cup of bread crumbs for the topping. How many cups of bread crumbs does Timothy need? (Lesson 6.1)

   (A) $\frac{1}{5}$ cup

   (B) $\frac{1}{3}$ cup

   (C) $\frac{2}{5}$ cup

   (D) $\frac{5}{6}$ cup

6. Jessie bought 3 T-shirts for $6 each and 4 T-shirts for $5 each. Which expression can you use to describe what Jessie bought? (Lesson 1.10)

   (A) $3 + 6 + 4 + 5$

   (B) $(3 + 6) \times (4 + 5)$

   (C) $(3 \times 6) + (4 \times 5)$

   (D) $(3 \times 6) \times (4 \times 5)$

Line Graphs

Use the table for 1–5.

Hourly Temperature							
Time	10 A.M.	11 A.M.	12 noon	1 P.M.	2 P.M.	3 P.M.	4 P.M.
Temperature (°F)	8	11	16	27	31	38	41

1. Write the related number pairs for the hourly temperature as ordered pairs.

   (10, 8); _____

   _____

2. What scale would be appropriate to graph the data?

   _____

3. What interval would be appropriate to graph the data?

   _____

4. Make a line graph of the data.

   _____

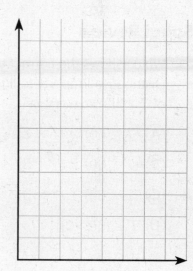

5. Use the graph to find the difference in temperature between 11 A.M. and 1 P.M.

   _____

Problem Solving REAL WORLD

6. Between which two hours did the least change in temperature occur?

   _____

   _____

7. What was the change in temperature between 12 noon and 4 P.M.?

   _____

Lesson Check

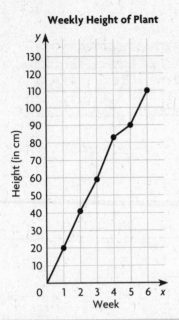

Weekly Height of Plant

1. How many centimeters did the plant grow in the first three weeks?

   (A) 20 cm   (C) 59 cm

   (B) 41 cm   (D) 83 cm

2. Between which two weeks did the plant grow the least?

   (A) Weeks 2 and 3

   (B) Weeks 3 and 4

   (C) Weeks 4 and 5

   (D) Weeks 5 and 6

Spiral Review

3. Which shows the correct use of the Distributive Property to find the product of 7×63? **(Lesson 1.10)**

   (A) $(7 \times 60) \times (7 \times 3)$

   (B) $(7 + 60) \times (7 + 3)$

   (C) $(7 \times 60) + (7 \times 3)$

   (D) $7 + (60 \times 3)$

4. Ali multiplies 3 numbers using the expressions $a \times (b \times c)$ and $(a \times b) \times c$. Which property of multiplication does Ali use? **(Lesson 1.3)**

   (A) Associative Property of Multiplication

   (B) Commutative Property of Multiplication

   (C) Distributive Property of Multiplication

   (D) Identity Property of Multiplication

5. A student athlete runs $3\frac{1}{3}$ miles in 30 minutes. A professional runner can run $1\frac{1}{4}$ times as far in 30 minutes. How far can the professional runner run in 30 minutes?

   (Lesson 7.9)

   (A) $3\frac{1}{12}$ miles   (C) $4\frac{2}{7}$ miles

   (B) $4\frac{1}{6}$ miles   (D) $4\frac{7}{12}$ miles

6. A recipe for salad dressing calls for $\frac{1}{4}$ cup of vinegar. You have 4 cups of vinegar. How many batches of salad dressing could you make with the vinegar? **(Lesson 8.4)**

   (A) 1

   (B) 4

   (C) 8

   (D) 16

Name _____

Numerical Patterns

Complete the rule that describes how one sequence is related to the other. Use the rule to find the unknown term.

1. Multiply the number of laps by __**50**__ to find the number of yards.

   Think: The number of yards is 50 times the number of laps.

Swimmers	1	2	3	4
Number of Laps	4	8	12	16
Number of Yards	200	400	600	**800**

2. Multiply the number of pounds by _____ to find total cost.

Boxes	1	2	3	4	6
Number of Pounds	3	6	9	12	18
Total Cost ($)	12	24	36	48	

3. Multiply the number of hours by _____ to find the number of miles.

Cars	1	2	3	4
Number of Hours	2	4	6	8
Number of Miles	130	260	390	

4. Multiply the number of hours by _____ to find the amount earned.

Days	1	2	3	4	7
Number of Hours	8	16	24	32	56
Amount Earned ($)	96	192	288	384	

Problem Solving REAL WORLD

5. A map distance of 5 inches represents 200 miles of actual distance. Suppose the distance between two cities on the map is 7 inches. What is the actual distance between the two cities? Write the rule you used to find the actual distance.

   _____

   _____

   _____

6. To make one costume, Rachel uses 6 yards of material and 3 yards of trim. Suppose she uses a total of 48 yards of material to make several costumes. How many yards of trim does she use? Write the rule you used to find the number of yards of trim.

   _____

   _____

   _____

Lesson Check

Use the table below to answer questions 1 and 2.

Term Number	1	2	3	4	6
Sequence 1	4	8	12	16	24
Sequence 2	12	24	36	48	?

1. What rule could you write that relates Sequence 2 to Sequence 1?

   (A) Add 8.

   (B) Multiply by 3.

   (C) Multiply by 4.

   (D) Add 48.

2. What is the unknown number in Sequence 2?

   (A) 48   (C) 72

   (B) 60   (D) 96

Spiral Review

3. What is the value of the following expression? **(Lesson 1.12)**

$$40 - (3 + 2) \times 6$$

   (A) 10

   (B) 49

   (C) 210

   (D) 234

4. What is the value of the digit 9 in the number 597,184? **(Lesson 1.2)**

   (A)   900

   (B)   9,000

   (C)   90,000

   (D)   900,000

5. Which is the best estimate for the sum of $\frac{3}{8}$ and $\frac{1}{12}$? **(Lesson 6.3)**

   (A) 0

   (B) $\frac{1}{2}$

   (C) 1

   (D) 4

6. Terry uses 3 cups of pecans to decorate the tops of 12 pecan pies. She puts an equal amount of pecans on each pie. How many cups of pecans does she put on each pie? **(Lesson 8.3)**

   (A) 9 cups

   (B) 4 cups

   (C) $\frac{1}{3}$ cup

   (D) $\frac{1}{4}$ cup

Name _____

Problem Solving • Find a Rule

Write a rule and complete the table. Then answer the question.

1. Faye buys 15 T-shirts, which are on sale for $3 each. How much money does Faye spend?

Number of T-Shirts	1	2	3	5	10	15
Amount Spent ($)	3	6	9			

Possible rule:

Multiply the number

of T-shirts by 3.

The total amount Faye spends is ___$45___.

2. The Gilman family joins a fitness center. They pay $35 per month. By the 12th month, how much money will the Gilman family have spent?

Number of Months	1	2	3	4	5	12
Total Amount of Money Spent ($)	35	70				

Possible rule:

_____

_____

The Gilman family will have spent _____.

3. Hettie is stacking paper cups. Each stack of 15 cups is 6 inches high. What is the total height of 10 stacks of cups?

Number of stacks	1	2	3	10
Height (in.)	6	12	18	

Possible rule:

_____

_____

The total height of 10 stacks is _____.

Lesson Check

1. How many squares are needed to make the eighth figure in the pattern?

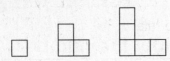

   Ⓐ 14

   Ⓑ 15

   Ⓒ 16

   Ⓓ 17

2. Which expression could describe the number of squares in the next figure in the pattern, Figure 4?

  Figure 1   Figure 2   Figure 3

  2 squares   5 squares   8 squares

   Ⓐ 6 + 2   Ⓒ 8 + 3

   Ⓑ 6 + 3   Ⓓ 8 + 4

Spiral Review

3. A bakery displays their cookies equally on 7 trays. If there are 567 cookies, how many cookies are on each tray? (Lesson 2.2)

   Ⓐ 487

   Ⓑ 486

   Ⓒ 81

   Ⓓ 80

4. Ms. Angelino made 2 pans of lasagna and cut each pan into twelfths. Her family ate $1\frac{1}{12}$ pans of lasagna for dinner. How many pans of lasagna were left? (Lesson 6.7)

   Ⓐ $\frac{11}{12}$

   Ⓑ $1\frac{11}{12}$

   Ⓒ $2\frac{1}{12}$

   Ⓓ $3\frac{1}{12}$

5. What is the next number in this pattern? (Lesson 3.10)

$$0.54, 0.6, 0.66, 0.72, \blacksquare, \ldots$$

   Ⓐ 0.76

   Ⓑ 0.78

   Ⓒ 0.8

   Ⓓ 0.82

6. How do you write 100 as a power of 10? (Lesson 1.4)

   Ⓐ 10^0

   Ⓑ 10^1

   Ⓒ 10^2

   Ⓓ 10^3

Name _____

Graph and Analyze Relationships

Graph and label the related number pairs as ordered pairs.
Then complete and use the rule to find the unknown term.

1. Multiply the number of yards by __**3**__ to find
   the number of feet.

Yards	1	2	3	4
Feet	3	6	9	**12**

2. Multiply the number of quarts by _____ to
   find the number of cups that measure the
   same amount.

Quarts	1	2	3	4	5
Cups	4	8	12	16	

Problem Solving REAL WORLD

3. How can you use the graph for Exercise 2 to find how
   many cups are in 9 quarts?

_____

_____

_____

_____

_____

4. How many cups are equal to 9 quarts? _____

Lesson Check

Use the data to complete the graph. Then answer the questions.

Paola is making a pitcher of iced tea. For each cup of water, she uses 3 tablespoons of powdered iced tea mix.

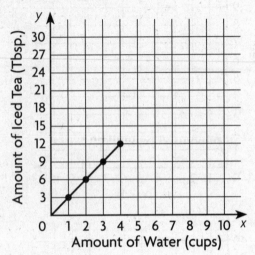

1. What rule relates the amount of iced tea mix to the amount of water?

   Ⓐ Multiply the amount of mix by 6.

   Ⓑ Multiply the amount of mix by 3.

   Ⓒ Multiply the amount of mix by $\frac{1}{3}$.

   Ⓓ Multiply the amount of mix by $\frac{1}{6}$.

2. Suppose Paola uses 18 tablespoons of iced tea mix. How many cups of water does she need to use?

   Ⓐ 3 cups

   Ⓑ 6 cups

   Ⓒ 9 cups

   Ⓓ 54 cups

Spiral Review

3. A biologist counted 10,000 migrating monarch butterflies. How do you express 10,000 as a power of 10? **(Lesson 1.4)**

   Ⓐ 10^2

   Ⓑ 10^3

   Ⓒ 10^4

   Ⓓ 10^5

4. For which expression will the quotient be greater than 100? **(Lesson 2.6)**

   Ⓐ $5,394 \div 57$

   Ⓑ $6,710 \div 69$

   Ⓒ $7,198 \div 74$

   Ⓓ $8,426 \div 82$

5. What is $54.38 + 29.7$? **(Lesson 3.8)**

   Ⓐ 57.35

   Ⓑ 83.45

   Ⓒ 83.08

   Ⓓ 84.08

6. On a certain day, $1 is worth 30.23 Russian rubles. Omar has $75. How many rubles will he get in exchange? **(Lesson 4.5)**

   Ⓐ 2,267.25

   Ⓑ 2,256.25

   Ⓒ   362.76

   Ⓓ   2.48

Name _____

Chapter 9 Extra Practice

Lesson 9.1

Use the data to complete the line plot. Then answer the questions.

1. Annabelle measured and recorded the amount of rainfall for 9 days in a row. Her results are shown below.

   $\frac{3}{10}$ in., $\frac{1}{5}$ in., $\frac{1}{10}$ in., $\frac{1}{10}$ in., $\frac{3}{10}$ in.,
   $\frac{2}{5}$ in., $\frac{1}{5}$ in., $\frac{1}{5}$ in., $\frac{1}{5}$ in.

2. What was the average amount of rainfall per day for the 9-day period?

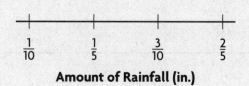

Amount of Rainfall (in.)

_____

Lesson 9.2

Use the coordinate grid to write an ordered pair for the given point.

1. A

   _____

2. F

   _____

3. L

   _____

4. C

   _____

5. N

   _____

6. W

   _____

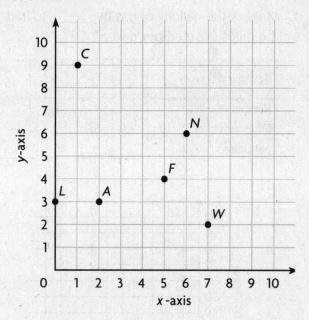

Plot and label the points on the coordinate grid.

7. P (6, 0)   8. T (2, 8)   9. M (8, 10)

10. R (9, 0)   11. Q (0, 0)   12. D (2, 1)

Lessons 9.3 - 9.4

Use the table for 1–2.

1. Make a line graph of the data.

Afternoon Temperature				
Time	1 P.M.	2 P.M.	3 P.M.	4 P.M.
Temperature (in °C)	15	25	20	12

2. Use the graph to find the difference in temperature between 2 P.M. and 3 P.M.

_____

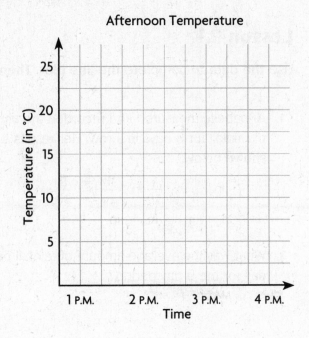

Afternoon Temperature

Lessons 9.5 - 9.7

Graph and label the related number pairs as ordered pairs. Then complete and use the rule to find the unknown term.

1. Multiply the gallons of gas by _____ to find the miles traveled per gallon.

Gallons of Gas	1	2	3	4	5
Number of Miles Traveled per Gallon	19	38	57		

2. Multiply the minutes by _____ to find the total number of seconds.

Minutes	1	2	3	5
Seconds	60	120	180	

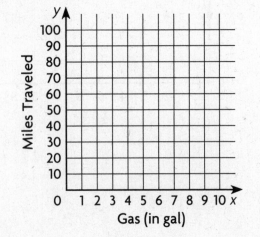

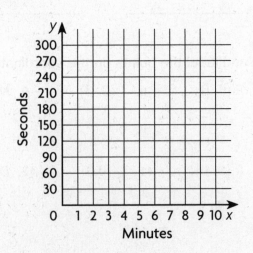

School-Home Letter

© Houghton Mifflin Harcourt Publishing Company

Dear Family,

Throughout the next few weeks, our math class will be learning how to compare and convert measurements. The students will use appropriate customary and metric units and conversion tables.

You can expect to see homework that includes comparing and converting length, weight/mass, capacity, and time.

Here is a sample of how your child will be taught to convert and compare weight.

Vocabulary

capacity The amount a container can hold when filled

elapsed time The amount of time that passes between the start of an activity and the end of that activity

gram A metric unit of mass

mass The amount of matter in an object

pound A customary unit of weight; 1 pound = 16 ounces

weight The measure of how heavy something is

🔒 MODEL Customary Weight

Convert 2 pounds to ounces. Compare the converted measure to 30 ounces.

STEP 1

1 pound is equal to 16 ounces.

$$\underset{\displaystyle 2}{\text{total pounds}} \times \underset{\displaystyle 16}{\text{ounces in 1 pound}} = \underset{\displaystyle 32}{\text{total ounces}}$$

STEP 2

Compare. Write <, >, or =.

32 ounces ◯ 30 ounces

32 > 30

Tips

Converting Units of Measure

Draw a picture to understand how units are related. When converting from a larger unit to a smaller unit, multiply. When converting from a smaller unit to a larger unit, divide.

Activity

Encourage your child to commit most of the unit conversions to memory. It will be useful for years to come. You can make a series of flash cards with equivalent measures on either side of the card, and work together to practice with unit conversions each night.

Carta
para la casa

Querida familia,

Durante las próximas semanas, en la clase de matemáticas aprenderemos a comparar y convertir medidas. Usaremos las unidades de adecuadas los sistemas usual y métrico, y tablas de conversión.

Llevaré a la casa tareas con actividades para comparar y convertir medidas de longitud, peso/masa, capacidad y tiempo.

Este es un ejemplo de la manera como aprenderemos a convertir y comparar medidas de peso.

Vocabulario

capacidad La cantidad que le cabe a un recipiente cuando se llena

tiempo transcurrido La cantidad de tiempo que pasa entre el comienzo y el final de una actividad

gramo Una unidad métrica de masa

masa La cantidad de materia que tiene un objeto

libra Una unidad usual de peso; 1 libra = 16 onzas

peso La medida de qué tan pesado es algo

🔒 MODELO El peso en el sistema usual

Convierte 2 libras a onzas. Compara la medida que convertiste con 30 onzas.

PASO 1

1 libra es igual a 16 onzas.

total de libras	×	onzas en 1 libra	=	total de onzas
↓		↓		↓
2		16		32

PASO 2

Compara. Escribe $<$, $>$ o $=$.

32 onzas ◯ 30 onzas

$32 > 30$

Pistas

Convertir unidades de medida

Haz un dibujo para entender cómo se relacionan las unidades. Cuando conviertas de una unidad mayor a una menor, multiplica. Cuando conviertas de una unidad menor a una mayor, divide.

Actividad

Anime a su hijo o hija a memorizar la mayoría de las conversiones de unidades. Es algo que le será útil en el futuro. Puede crear una serie de tarjetas nemotécnicas con medidas equivalentes en los dos lados de cada tarjeta, trabajen juntos y practiquen las conversiones de unidades en la noche.

Name _____

Customary Length

Convert.

1. 12 yd = __**36**__ ft

total yards   feet in 1 yard   total feet

12   ×   3   =   36

12 yards = 36 feet

2. 5 ft = _____ in.

3. 5 mi = _____ ft

4. 240 in. = _____ ft

5. 100 yd = _____ ft

6. 10 ft = _____ in.

7. 150 in. = _____ ft _____ in.

8. 7 yd 2 ft = _____ ft

9. 10 mi = _____ ft

Compare. Write <, >, or =.

10. 23 in. ◯ 2 ft

11. 25 yd ◯ 75 ft

12. 6,200 ft ◯ 1 mi 900 ft

13. 100 in. ◯ 3 yd 1 ft

14. 1,000 ft ◯ 300 yd

15. 500 in. ◯ 40 ft

Problem Solving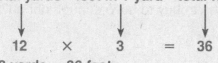

16. Marita orders 12 yards of material to make banners. If she needs 1 foot of fabric for each banner, how many banners can she make?

_____

17. Christy bought an 8-foot piece of lumber to trim a bookshelf. Altogether, she needs 100 inches of lumber for the trim. Did Christy buy enough lumber? Explain.

_____

_____

Lesson Check

1. Jenna's garden is 5 yards long. How long is her garden in feet?

   (A) 60 feet

   (B) 15 feet

   (C) 8 feet

   (D) 2 feet

2. Ellen needs to buy 180 inches of ribbon to wrap a large present. The store sells ribbon only in whole yards. How many yards does Ellen need to buy to have enough ribbon?

   (A) 3 yards

   (B) 4 yards

   (C) 5 yards

   (D) 6 yards

Spiral Review

3. McKenzie works for a catering company. She is making iced tea for an upcoming event. For each container of tea, she uses 16 tea bags and 3 cups of sugar. If McKenzie uses 64 tea bags, how many cups of sugar will she use? **(Lesson 9.6)**

   (A) $\frac{3}{4}$ cup

   (B) 4 cups

   (C) 8 cups

   (D) 12 cups

4. Javier bought 48 sports cards at a yard sale. Of the cards, $\frac{3}{8}$ were baseball cards. How many cards were baseball cards? **(Lesson 7.1)**

   (A) 48

   (B) 18

   (C) 6

   (D) 3

5. Which is the quotient of 396 divided by 12?
   (Lesson 2.6)

   (A) 31

   (B) 33

   (C) 36

   (D) 38

6. What is the unknown number in Sequence 2 in the chart? What rule can you write that relates Sequence 2 to Sequence 1? **(Lesson 9.5)**

Sequence Number	1	2	3	8	10
Sequence 1	4	8	12	32	40
Sequence 2	8	16	24	64	?

   (A) 40; Multiply by 1.

   (B) 60; Add 20.

   (C) 80; Multiply by 2.

   (D) 20; Divide by 2.

Customary Capacity

Convert.

1. 5 gal = **40** pt

   Think: 1 gallon = 4 quarts

            1 quart = 2 pints

2. 192 fl oz = _____ pt

3. 15 pt = _____ c

4. 240 fl oz = _____ c

5. 32 qt = _____ gal

6. 10 qt = _____ c

7. 48 c = _____ qt

8. 72 pt = _____ gal

9. 128 fl oz = _____ pt

Compare. Write <, >, or =.

10. 17 qt ( ) 4 gal

11. 96 fl oz ( ) 8 pt

12. 400 pt ( ) 100 gal

13. 100 fl oz ( ) 16 pt

14. 74 fl oz ( ) 8 c

15. 12 c ( ) 3 qt

Problem Solving  REAL WORLD

16. Vickie made a recipe for 144 fluid ounces of scented candle wax. How many 1-cup candle molds can she fill with the recipe?

    _____

17. A recipe calls for 32 fluid ounces of heavy cream. How many 1-pint containers of heavy cream are needed to make the recipe?

    _____

Lesson Check

1. Rosa made 12 gallons of lemonade to sell at a lemonade stand. How many pints of lemonade did she make?

   (A) 96 pints

   (B) 48 pints

   (C) 3 pints

   (D) $1\frac{1}{2}$ pints

2. Ebonae's fish tank holds 40 gallons. How many quarts does the fish tank hold?

   (A) 4 quarts

   (B) 10 quarts

   (C) 80 quarts

   (D) 160 quarts

Spiral Review

3. A mountain climber climbed 15,840 feet on her way to the summit of a mountain. How many miles did she climb? (Lesson 10.1)

   (A) 1 mile

   (B) 2 miles

   (C) 3 miles

   (D) 4 miles

4. Jamal is making pancakes. He has $6\frac{3}{4}$ cups of batter, but he needs a total of 12 cups. How much more batter does Jamal need?
   (Lesson 6.6)

   (A) $5\frac{1}{4}$ cups

   (B) $5\frac{3}{4}$ cups

   (C) $6\frac{1}{4}$ cups

   (D) $18\frac{3}{4}$ cups

5. At a building site, there are 16 pallets with sacks of cement. The total weight of all the pallets and cement is 4,856 pounds. Each pallet with cement weighs the same amount. How much does each pallet with cement weigh? (Lesson 2.7)

   (A) 304 pounds

   (B) $303\frac{1}{2}$ pounds

   (C) 303 pounds

   (D) 300 pounds

6. A publisher shipped 15 boxes of books to a bookstore. Each box contained 32 books. How many books in all did the publisher ship to the bookstore? (Lesson 1.7)

   (A) 560

   (B) 480

   (C) 400

   (D) 320

Name _____

Weight

Convert.

1. 96 oz = ___**6**___ lb

total oz   oz in 1 lb   total lb

96   ÷   16   =   6

2. 6 T = _____ lb

3. 18 lb = _____ oz

4. 3,200 oz = _____ lb

5. 12 T = _____ lb

6. 9 lb = _____ oz

7. 7 lb = _____ oz

8. 100 lb = _____ oz

9. 60,000 lb = _____ T

Compare. Write <, >, or =.

10. 40 oz ◯ 4 lb

11. 80 oz ◯ 5 lb

12. 5,000 lb ◯ 5 T

13. 18,000 lb ◯ 9 T

14. 25 lb ◯ 350 oz

15. 27 oz ◯ 2 lb

Problem Solving REAL WORLD

16. Mr. Fields ordered 3 tons of gravel for a driveway at a factory. How many pounds of gravel did he order?

_____

17. Sara can take no more than 22 pounds of luggage on a trip. Her suitcase weighs 112 ounces. How many more pounds can she pack without going over the limit?

_____

Lesson Check

1. Paolo's puppy weighed 11 pounds at the vet's office. What is this weight in ounces?

    (A) 16 ounces

    (B) 32 ounces

    (C) 166 ounces

    (D) 176 ounces

2. The weight limit on a bridge is 5 tons. What is this weight in pounds?

    (A) 80 pounds

    (B) 5,000 pounds

    (C) 10,000 pounds

    (D) 20,000 pounds

Spiral Review

3. There are 20 guests at a party. The host has 8 gallons of punch. He estimates that each guest will drink 2 cups of punch. If his estimate is correct, how much punch will be left over at the end of the party? **(Lesson 10.2)**

    (A) 16 cups

    (B) 40 cups

    (C) 88 cups

    (D) 128 cups

4. A typical lap around a track in the United States has a length of 440 yards. How many laps would need to be completed to run a mile? **(Lesson 10.1)**

    (A) 4

    (B) 12

    (C) 40

    (D) 440

5. A recipe for sweet potato pie calls for $\frac{3}{4}$ cup of milk. Martina has 6 cups of milk. How many sweet potato pies can she make with that amount of milk? **(Lesson 8.4)**

    (A) 2

    (B) 4

    (C) 8

    (D) 16

6. Which of the following is the best estimate for the total weight of these cold meats: $1\frac{7}{8}$ pounds of bologna, $1\frac{1}{2}$ pounds of ham, and $\frac{7}{8}$ pound of roast beef? **(Lesson 6.6)**

    (A) 3 pounds

    (B) $3\frac{1}{2}$ pounds

    (C) 4 pounds

    (D) $4\frac{1}{2}$ pounds

Name _____

Multistep Measurement Problems

Solve.

1. A cable company has 5 miles of cable to install. How many 100-yard lengths of cable can be cut?

   Think: 1,760 yards = 1 mile.
   So the cable company has 5 × 1,760, or 8,800 yards of cable.

   Divide. 8,800 ÷ 100 = 88

   ## 88 lengths

2. Afton made a chicken dish for dinner. She added a 10-ounce package of vegetables and a 14-ounce package of rice to 40 ounces of chicken. What was the total weight of the chicken dish in pounds?

3. A jar contains 26 fluid ounces of spaghetti sauce. How many cups of spaghetti sauce do 4 jars contain?

4. Coach Kent brings 3 quarts of sports drink to soccer practice. He gives the same amount of the drink to each of his 16 players. How many ounces of the drink does each player get?

5. Leslie needs 324 inches of fringe to put around the edge of a tablecloth. The fringe comes in lengths of 10 yards. If Leslie buys 1 package of fringe, how many feet of fringe will she have left over?

6. Darnell rented a moving truck. The weight of the empty truck was 7,860 pounds. When Darnell filled the truck with his items, it weighed 6 tons. What was the weight in pounds of the items that Darnell placed in the truck?

Problem Solving REAL WORLD

7. A pitcher contains 40 fluid ounces of iced tea. Shelby pours 3 cups of iced tea. How many pints of iced tea are left in the pitcher?

8. Olivia ties 2.5 feet of ribbon onto one balloon. How many yards of ribbon does Olivia need for 18 balloons?

Lesson Check

1. Leah is buying curtains for her bedroom window. She wants the curtains to hang from the top of the window to the floor. The window is 4 feet high. The bottom of the window is $2\frac{1}{2}$ feet above the floor. What curtain length should Leah buy?

   (A) 72 inches

   (B) 78 inches

   (C) 84 inches

   (D) 104 inches

2. Brady buys 3 gallons of fertilizer for his lawn. After he finishes spraying the lawn, he has 1 quart of fertilizer left over. How many quarts of fertilizer did Brady spray on the lawn?

   (A) 3 quarts

   (B) 7 quarts

   (C) 11 quarts

   (D) 15 quarts

Spiral Review

3. A jump rope is 9 feet long. How long is the jump rope in yards? (Lesson 10.1)

   (A) $\frac{3}{4}$ yard

   (B) 3 yards

   (C) 27 yards

   (D) 108 yards

4. Which of the following measurements is NOT equal to 8 cups? (Lesson 10.2)

   (A) 1 gallon

   (B) 2 quarts

   (C) 4 pints

   (D) 64 fluid ounces

5. What is the unknown number in Sequence 2 in the chart? (Lesson 9.5)

Sequence Number	1	2	3	5	7
Sequence 1	3	6	9	15	21
Sequence 2	6	12	18	30	?

   (A) 32

   (B) 35

   (C) 36

   (D) 42

6. A farmer divides 20 acres of land into $\frac{1}{4}$-acre sections. Into how many sections does the farmer divide her land? (Lesson 8.2)

   (A) 4

   (B) 5

   (C) 16

   (D) 80

Name _____

Metric Measures

Convert.

1. 16 m = **16,000** mm

        number of   millimeters   number of
        meters   in 1 meter   millimeters

         16   ×   1,000   =   16,000
      16 m = 16,000 mm

2. 6,500 cL = _____ L

3. 15 cm = _____ mm

4. 3,200 g = _____ kg

5. 12 L = _____ mL

6. 200 cm = _____ m

7. 70,000 g = _____ kg

8. 100 dL = _____ L

9. 60 m = _____ mm

Compare. Write <, >, or =.

10. 900 cm ◯ 9,000 mm

11. 600 km ◯ 5 m

12. 5,000 cm ◯ 5 m

13. 18,000 g ◯ 10 kg

14. 8,456 mL ◯ 9 L

15. 2 m ◯ 275 cm

Problem Solving REAL WORLD

16. Bria ordered 145 centimeters of fabric. Jayleen ordered 1.5 meters of fabric. Who ordered more fabric?

17. Ed fills his sports bottle with 1.2 liters of water. After his bike ride, he drinks 200 milliliters of the water. How much water is left in Ed's sports bottle?

_____

_____

Lesson Check

1. Quan bought 8.6 meters of fabric. How many centimeters of fabric did he buy?

   Ⓐ   86 centimeters

   Ⓑ   860 centimeters

   Ⓒ   8,600 centimeters

   Ⓓ 86,000 centimeters

2. Jason takes 2 centiliters of medicine. How many milliliters is this?

   Ⓐ 200 milliliters

   Ⓑ   20 milliliters

   Ⓒ   0.2 milliliter

   Ⓓ   0.02 milliliter

Spiral Review

3. Yolanda needs 5 pounds of ground beef to make lasagna for a family reunion. One package of ground beef weighs $2\frac{1}{2}$ pounds. Another package weighs $2\frac{3}{5}$ pounds. How much ground beef will Yolanda have left over after making the lasagna? **(Lesson 6.6)**

   Ⓐ $\frac{1}{2}$ pound

   Ⓑ $\frac{1}{3}$ pound

   Ⓒ $\frac{1}{5}$ pound

   Ⓓ $\frac{1}{10}$ pound

4. A soup recipe calls for $2\frac{3}{4}$ quarts of vegetable broth. An open can of broth contains $\frac{1}{2}$ quart of broth. How much more broth do you need to make the soup? **(Lesson 6.6)**

   Ⓐ $\frac{1}{2}$ quart

   Ⓑ 2 quarts

   Ⓒ $2\frac{1}{4}$ quarts

   Ⓓ $3\frac{1}{4}$ quarts

5. Which point on the graph is located at $(4, 2)$? **(Lesson 9.2)**

   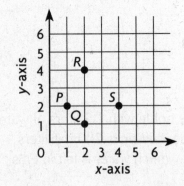

   Ⓐ   P   Ⓒ   R

   Ⓑ   Q   Ⓓ   S

6. A bakery supplier receives an order for 2 tons of sugar from a bakery chain. The sugar is shipped in crates. Each crate holds eight 10-pound bags of sugar. How many crates does the supplier need to ship to fulfill the order? **(Lesson 10.4)**

   Ⓐ   50

   Ⓑ   80

   Ⓒ   200

   Ⓓ 4,000

Name _____

Problem Solving • Customary and Metric Conversions

Solve each problem by making a table.

1. Thomas is making soup. His soup pot holds 8 quarts of soup. How many 1-cup servings of soup will Thomas make?

Number of Quarts	1	2	3	4	8
Number of Cups	4	8	12	16	32

32 1-cup servings

2. Paulina works out with a 2.5-kilogram mass. What is the mass of the 2.5-kilogram mass in grams?

3. Alex lives 500 yards from the park. How many inches does Alex live from the park?

4. Emma uses a 250-meter roll of crepe paper to make streamers. How many dekameters of crepe paper does Emma use?

5. A flatbed truck is loaded with 7,000 pounds of bricks. How many tons of brick are on the truck?

Lesson Check

1. At the hairdresser, Jenny had 27 centimeters cut off her hair. How many decimeters of hair did Jenny have cut off?

   Ⓐ 0.027 dm

   Ⓑ 0.27 dm

   Ⓒ 2.7 dm

   Ⓓ 270 dm

2. Marcus needs 108 inches of wood to make a frame. How many feet of wood does Marcus need for the frame?

   Ⓐ 3 feet

   Ⓑ 6 feet

   Ⓒ $7\frac{1}{2}$ feet

   Ⓓ 9 feet

Spiral Review

3. Tara lives 35,000 meters from her grandparents. How many kilometers does Tara live from her grandparents? **(Lesson 10.5)**

   Ⓐ 3.5 km

   Ⓑ 35 km

   Ⓒ 350 km

   Ⓓ 3,500 km

4. Dane's puppy weighed 8 ounces when it was born. Now the puppy weighs 18 times as much as it did when it was born. How many pounds does Dane's puppy weigh now?
   (Lesson 10.4)

   Ⓐ 9 pounds

   Ⓑ 12 pounds

   Ⓒ 16 pounds

   Ⓓ 18 pounds

5. A carpenter is cutting dowels from a piece of wood that is 10 inches long. How many $\frac{1}{2}$-inch dowels can the carpenter cut?
   (Lesson 8.4)

   Ⓐ 2

   Ⓑ 5

   Ⓒ 15

   Ⓓ 20

6. Which ordered pair describes the location of point X? **(Lesson 9.2)**

   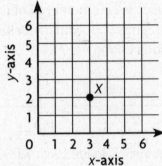

   Ⓐ (2, 3)

   Ⓑ (2, 2)

   Ⓒ (3, 2)

   Ⓓ (3, 3)

Elapsed Time

Convert.

1. 5 days = ____**120**____ hr

2. 8 hr = _____ min

3. 30 min = _____ sec

Think: 1 day = 24 hours
  5 × 24 = 120

4. 15 hr = _____ min

5. 5 yr = _____ d

6. 7 d = _____ hr

7. 24 hr = _____ min

8. 600 sec = _____ min

9. 60,000 min = _____ hr

Find the start, elapsed, or end time.

10. Start time: 11:00 A.M.

   Elapsed time: 4 hours 5 minutes

   End time: _____

11. Start time: 6:30 P.M.

   Elapsed time: 2 hours 18 minutes

   End time: _____

12. Start time: _____

   Elapsed time: $9\frac{3}{4}$ hours

   End time: 6:00 P.M.

13. Start time: 2:00 P.M.

   Elapsed time: _____

   End time: 8:30 P.M.

Problem Solving REAL WORLD

14. Kiera's dance class starts at 4:30 P.M. and ends at 6:15 P.M. How long is her dance class?

   _____

15. Julio watched a movie that started at 11:30 A.M. and ended at 2:12 P.M. How long was the movie?

   _____

Lesson Check

1. Michelle went on a hike. She started on the trail at 6:45 A.M. and returned at 3:28 P.M. How long did she hike?

   (A) 3 hours 27 minutes

   (B) 4 hours 43 minutes

   (C) 6 hours 27 minutes

   (D) 8 hours 43 minutes

2. Grant started a marathon at 8:00 A.M. He took 4 hours 49 minutes to complete the marathon. When did he cross the finish line?

   (A) 12:11 P.M.

   (B) 12:49 P.M.

   (C) 2:11 P.M.

   (D) 2:49 P.M.

Spiral Review

3. Molly is filling a pitcher that holds 2 gallons of water. She is filling the pitcher with a 1-cup measuring cup. How many times will she have to fill the 1-cup measuring cup to fill the pitcher? **(Lesson 10.6)**

   (A) 4

   (B) 8

   (C) 16

   (D) 32

4. Which decimal is between 1.5 and 1.7?
   (Lesson 3.3)

   (A) 1.25

   (B) 1.625

   (C) 1.75

   (D) 1.83

5. Adrian's recipe for raisin muffins calls for $1\frac{3}{4}$ cups raisins for one batch of muffins. Adrian wants to make $2\frac{1}{2}$ batches of the muffins for a bake sale. How many cups of raisins will Adrian use? **(Lesson 7.9)**

   (A) $2\frac{1}{2}$ cups

   (B) $4\frac{1}{4}$ cups

   (C) $4\frac{3}{8}$ cups

   (D) $8\frac{3}{4}$ cups

6. Kevin is riding his bike on a $10\frac{1}{8}$-mile bike path. He has covered the first $5\frac{3}{4}$ miles already. How many miles does he have left to ride? **(Lesson 6.7)**

   (A) $4\frac{3}{8}$ miles

   (B) $4\frac{5}{8}$ miles

   (C) $5\frac{3}{8}$ miles

   (D) $5\frac{5}{8}$ miles

Chapter 10 Extra Practice

Lessons 10.1 - 10.3, 10.5

Convert.

1. 8 yd = _____ ft

2. 185 in. = _____ ft _____ in.

3. 2 mi = _____ ft

4. 8 c = _____ pt

5. 12 gal = _____ qt

6. 32 c = _____ fl oz

7. 6,000 lb = _____ T

8. 9 lb = _____ oz

9. 112 oz = _____ lb

10. 380 dm = _____ m

11. 90.51 L = _____ cL

12. 450 mg = _____ g

Compare. Write <, >, or =.

13. 9 ft ◯ 4 yd

14. 4 mi ◯ 15,840 ft

15. 5 yd 1 ft ◯ 192 in.

16. 10 gal ◯ 60 qt

17. 480 fl oz ◯ 24 pt

18. 16 cups ◯ 1 gal

19. 18 T ◯ 36,000 lb

20. 145 oz ◯ 9 lb

21. 1 T ◯ 3,400 lb

22. 45 hg ◯ 4.5 kg

23. 770 m ◯ 7 km

24. 875 cL ◯ 875 mL

Lesson 10.4

Solve.

1. An office supply company is shipping a case of paper to a school. There are 10 reams of paper in the case. If each ream of paper weighs 32 ounces, what is the weight, in pounds, of the case of paper?

_____

2. An adult blue whale weighs 120 tons. A baby blue whale weighs $\frac{1}{40}$ of the weight of the adult blue whale. How many pounds does the baby blue whale weigh?

_____

Lesson 10.6

Solve.

1. Kat has a smoothie company. She needs to make 240 cups of fruit smoothies for the next day. If she wants to store the smoothies overnight in quart containers, how many quart containers will Kat need?

_____

2. Ty needs to cut strips of wrapping paper that are each 9 inches wide. If Ty has 2 rolls of wrapping paper that are each 15 feet long, how many 9-inch strips can he cut?

_____

Lesson 10.7

Convert.

1. 2 yr = _____ d

2. 260 wk = _____ yr

3. 270 min = _____ hr _____ min

Find the start, elapsed, or end time.

4. Start time: _____

Elapsed time: $\frac{3}{4}$ hour

End time: 10:30 A.M.

5. Start time: 7:30 P.M.

Elapsed time: _____

End time: 9:29 P.M.

Vocabulary

congruent Having the same size and shape

trapezoid A quadrilateral with exactly one pair of parallel sides

polyhedron A three-dimensional figure with faces that are polygons

lateral faces Faces of a polyhedron that connect the bases

Dear Family,

Throughout the next few weeks, our math class will be studying two-dimensional and three-dimensional figures. The students will use definitions to identify and describe characteristics of these figures. We will also learn how to find volume of rectangular prisms.

You can expect to see homework that includes identifying types of triangles and quadrilaterals.

Here is a sample of how your child will be taught to classify a triangle by the length of its sides.

🔑 **MODEL** **Classify a triangle by the length of its sides.**

A triangle has side lengths 3 in., 2 in., and 3 in. What type of triangle is it?

STEP 1

Identify how many sides are congruent.

There are ___**2**___ sides with lengths of

___**3 in.**___

STEP 2

Determine the correct classification.

A triangle with two congruent sides is

___**isosceles**___.

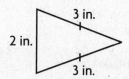

Tips

Congruent Figures

Congruent figures are figures that have the same size and shape.

If measurements aren't given and you need to check whether a figure has pairs of congruent sides or angles, trace the figure and cut out the tracing. Then fold the figure to see if the sides or angles match.

Activity

Try to have students commit most of the classifications of triangles, quadrilaterals, and polyhedrons to memory. You can make a series of flash cards with the classifications on one side of the card and definitions and/or sketches of examples on the other side of the card.

Carta para la casa

© Houghton Mifflin Harcourt Publishing Company

Vocabulario

congruentes Figuras que tienen el mismo tamaño y la misma forma

trapecio Un cuadrilátero que tiene exactamente 1 par de lados paralelos

poliedro Una figura tridimensional con caras que son polígonos

caras laterales Las caras poligonales de un poliedro que conectan las bases

Querida familia

Durante las próximas semanas, en la clase de matemáticas estudiaremos las figuras bidimensionales y tridimensionales. Usaremos las definiciones para identificar y describir las características de esas figuras. También aprenderemos a hallar el volumen de los prismas rectangulares.

Llevaré a la casa tareas con actividades para identificar diferentes tipos de triángulos y cuadriláteros.

Este es un ejemplo de la manera como aprenderemos a clasificar un triángulo por sus lados.

🔑 MODELO Clasificar un triángulo por sus lados.

Los lados de un triángulo miden 3 pulg., 2 pulg. y 3 pulg. ¿Qué tipo de triángulo es?

PASO 1

Identifica cuántos lados son iguales.

Hay **dos** lados que tienen la misma longitud de **3 pulg.**

PASO 2

Determina la clasificación correcta.

Un triángulo con dos lados congruentes es

isósceles

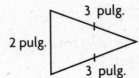

3 pulg.

2 pulg.

3 pulg.

Pistas

Figuras congruentes

Las figuras congruentes son figuras que tienen el mismo tamaño y la misma forma. Para comprobar si una figura tiene pares de lados o ángulos congruentes, dibuja la figura y recórtala. Luego dobla la figura para ver si los lados o los ángulos coinciden.

Para estar seguro de que dos figuras son congruentes, haz una lista de todos los lados y ángulos que corresponden uno con el otro y luego verifica que las medidas de cada par sean iguales.

Actividad

Anime a su hijo/a a memorizar las clasificaciones de los triángulos, los cuadriláteros y los poliedros. Puede hacer tarjetas nemotécnicas con las clasificaciones en un lado y las definiciones y/o ejemplos visuales en el otro lado de cada tarjeta.

Polygons

Name each polygon. Then tell whether it is a *regular polygon*
or *not a regular polygon.*

1.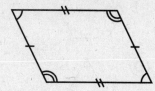

4 sides, 4 vertices, 4 angles means it is a

<u>**quadrilateral**</u>. The sides are

not all congruent, so it is <u>**not regular**</u>.

2.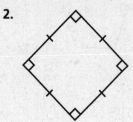

_____

3.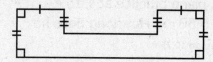

_____

4.

_____

5.

_____

6.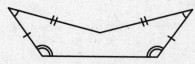

_____

Problem Solving

7. Sketch nine points. Then, connect the points
to form a closed plane figure. What kind of
polygon did you draw?

_____

8. Sketch seven points. Then, connect the points
to form a closed plane figure. What kind of
polygon did you draw?

_____

Lesson Check

1. Which of the following is a regular pentagon?

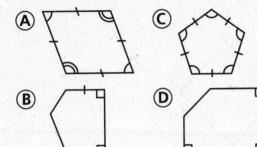

Ⓐ   Ⓒ

Ⓑ   Ⓓ

2. Which of the following is NOT a regular polygon?

Ⓐ   Ⓒ

Ⓑ   Ⓓ

Spiral Review

3. Ann needs 42 feet of fabric to make a small quilt. How many yards of fabric should she buy? **(Lesson 10.1)**

Ⓐ   13 yards

Ⓑ   14 yards

Ⓒ   21 yards

Ⓓ   126 yards

4. Todd begins piano practice at 4:15 P.M. and ends at 5:50 P.M. How long does he practice? **(Lesson 10.7)**

Ⓐ 25 minutes

Ⓑ 35 minutes

Ⓒ 1 hour 25 minutes

Ⓓ 1 hour 35 minutes

5. Jenna is organizing her barrettes into 6 boxes. She puts the same number of barrettes in each box. If Jenna has 30 barrettes, which expression can you use to find the number of barrettes in each box? **(Lesson 1.10)**

Ⓐ 6 × 30

Ⓑ 30 + 6

Ⓒ 30 − 6

Ⓓ 30 ÷ 6

6. Melody had $45. She spent $32.75 on a blouse. Then her mother gave her $15.50. How much money does Melody have now? **(Lesson 3.11)**

Ⓐ $12.25

Ⓑ $27.75

Ⓒ $48.25

Ⓓ $60.50

Triangles

Classify each triangle. Write *isosceles*, *scalene*, or *equilateral*.
Then write *acute*, *obtuse*, or *right*.

1.
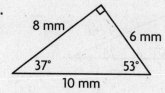

2.
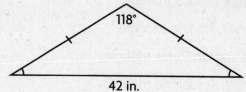

None of the side measures are equal. So, it is

___scalene___. There is a right

angle, so it is a _____right_____ triangle.

_____ _____

3.
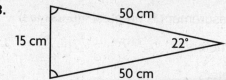

4.
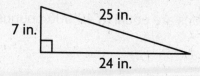

_____ _____

_____ _____

A triangle has sides with the lengths and angle measures given.
Classify each triangle. Write *scalene*, *isosceles*, or *equilateral*. Then
write *acute*, *obtuse*, or *right*.

5. sides: 44 mm, 28 mm, 24 mm

angles: 110°, 40°, 30°

6. sides: 23 mm, 20 mm, 13 mm

angles: 62°, 72°, 46°

_____ _____ _____ _____

Problem Solving

7. Mary says the pen for her horse is an acute right triangle. Is this possible? **Explain.**

8. Karen says every equilateral triangle is acute. Is this true? **Explain.**

_____ _____

Lesson Check

1. Which of the following triangles is impossible to draw?

   Ⓐ right obtuse triangle

   Ⓑ right scalene triangle

   Ⓒ acute isosceles triangle

   Ⓓ obtuse scalene triangle

2. What is the classification of the following triangle?

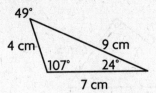

   Ⓐ scalene   Ⓒ isosceles

   Ⓑ right   Ⓓ acute

Spiral Review

3. How many tons are equal to 40,000 pounds? **(Lesson 10.3)**

   Ⓐ 2 tons

   Ⓑ 4 tons

   Ⓒ 20 tons

   Ⓓ 40 tons

4. Which measurement is greatest? **(Lesson 10.5)**

   Ⓐ 6 kilometers

   Ⓑ 60 meters

   Ⓒ 600 centimeters

   Ⓓ 6,000 millimeters

5. Which polygon is shown? **(Lesson 11.1)**

   Ⓐ quadrilateral

   Ⓑ pentagon

   Ⓒ hexagon

   Ⓓ octagon

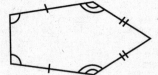

6. Which of the following is a regular polygon? **(Lesson 11.1)**

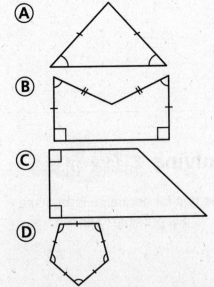

Name _____

Quadrilaterals

Classify the quadrilateral in as many ways as possible.
Write *quadrilateral, parallelogram, rectangle, rhombus, square,* **or** *trapezoid.*

1.
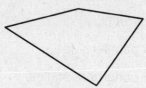

It has 4 sides, so it is a **quadrilateral** _____.
None of the sides are parallel, so there is

no other classification. _____.

2.
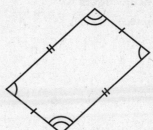

_____

3.
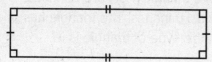

_____

4.
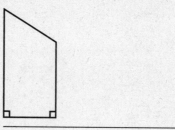

_____

5.
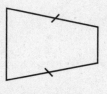

_____

6.

_____

Problem Solving REAL WORLD

7. Kevin claims he can draw a trapezoid with three right angles. Is this possible? **Explain.**

_____

_____

_____

_____

8. "If a figure is a square, then it is a regular quadrilateral." Is this true or false? **Explain.**

_____

_____

_____

_____

Lesson Check

1. A certain parallelogram has two pairs of parallel sides. Based on this, which of the following classifications is NOT correct?

   (A) quadrilateral

   (B) rectangle

   (C) square

   (D) trapezoid

2. Which of the following is NOT always true about a rhombus?

   (A) All sides are congruent.

   (B) All angles are congruent.

   (C) It has two pairs of parallel sides.

   (D) It has two pairs of congruent angles.

Spiral Review

3. How many kilograms are equal to 5,000 grams? **(Lesson 10.5)**

   (A) 500 kilograms

   (B) 50 kilograms

   (C) 5 kilograms

   (D) 0.5 kilogram

4. The sides of a triangle measure 6 inches, 8 inches, and 10 inches. The triangle has one 90° angle. What type of triangle is it?

   (Lesson 11.2)

   (A) scalene right

   (B) isosceles right

   (C) equilateral right

   (D) scalene acute

5. A warehouse has 355 books to ship. Each shipping carton holds 14 books. How many cartons does the warehouse need to ship all of the books? **(Lesson 2.7)**

   (A) 5

   (B) 25

   (C) 26

   (D) 30

6. Which of the following statements is NOT true? **(Lesson 11.1)**

   (A) A polygon has the same number of angles as it has sides.

   (B) In a regular polygon, all sides are congruent, and all angles are congruent.

   (C) Some triangles are regular.

   (D) All heptagons have 6 vertices.

Name _____

Problem Solving • Properties of Two-Dimensional Figures

Solve each problem.

1. Marcel thinks that quadrilateral *ABCD* at the right has two pairs of congruent sides, but he does not have a ruler to measure the sides. How can he show that the quadrilateral has two pairs of congruent sides?

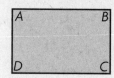

He can fold the quadrilateral in half both ways. If both sets of sides match, then they are congruent.

2. If what Marcel thinks about his quadrilateral is true, what type of quadrilateral does he have? _____

3. Richelle drew hexagon *KLMNOP* at the right. She thinks the hexagon has six congruent angles. How can she show that the angles are congruent without using a protractor to measure them?

_____

_____

4. Jerome drew a triangle with vertices *S*, *T*, and *U*. He thinks ∠*TSU* and ∠*TUS* are congruent. How can Jerome show that the angles are congruent without measuring the angles?

_____

_____

5. If Jerome is correct, what type of triangle did he draw?

_____

Lesson Check

1. Peter knows that pentagon *DEFGH* has 5 congruent sides. How can he determine if the pentagon has 5 congruent angles without measuring?

Ⓐ He can trace and fold the figure to check if the angles match up.

Ⓑ All pentagons have congruent angles.

Ⓒ If a polygon has an odd number of angles, the angles cannot be congruent.

Ⓓ He can use an index card to check that at least one angle is right.

2. Tina knows that quadrilateral *WXYZ* has 2 pairs of congruent angles. She thinks that all 4 sides look congruent but she does not have a ruler. How can Tina determine whether she is correct?

Ⓐ She can fold the figure to check that *WX* and *YZ* match up.

Ⓑ She can fold the figure to check that *XY* and *WZ* match up.

Ⓒ She can fold the figure along both diagonals, *XZ* and *WY*, to check if the sides match.

Ⓓ The sides cannot all be congruent if the angles are not all congruent.

Spiral Review

3. How many ounces are in 50 pounds?

(Lesson 10.3)

Ⓐ 800 ounces

Ⓑ 500 ounces

Ⓒ 400 ounces

Ⓓ 200 ounces

4. How many minutes are there in 40 hours?

(Lesson 10.7)

Ⓐ 4,000 minutes

Ⓑ 2,400 minutes

Ⓒ 960 minutes

Ⓓ 240 minutes

5. Which of the following angle measures could NOT represent an angle measure of an acute triangle? (Lesson 11.2)

Ⓐ 33°

Ⓑ 78°

Ⓒ 81°

Ⓓ 92°

6. Which of the following angle measures represents the measure of each of the four angles of a square? (Lesson 11.3)

Ⓐ 45°

Ⓑ 60°

Ⓒ 90°

Ⓓ 100°

Three-Dimensional Figures

Classify the solid figure. Write *prism, pyramid, cone, cylinder,* **or** *sphere.*

1.
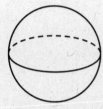

There are no bases. There is
1 curved surface. It is a

____sphere____.

2.
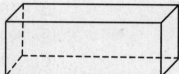

_____

3.

_____

Name the solid figure.

4.
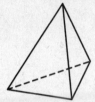

_____

5.
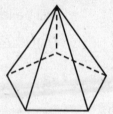

_____

6.
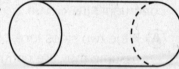

_____

7.
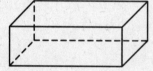

_____

8.
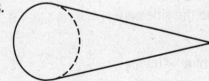

_____

9.
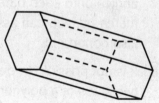

_____

Problem Solving REAL WORLD

10. Darrien is making a solid figure out of folded
    paper. His solid figure has six congruent faces
    that are all squares. What solid figure did
    Darrien make?

    _____

11. Nanako said she drew a square pyramid
    and that all of the faces are triangles. Is this
    possible? **Explain.**

    _____

Lesson Check

1. Luke made a model of a solid figure with 1 circular base and 1 curved surface. What solid figure did he make?

   A cone

   B cylinder

   C sphere

   D triangular pyramid

2. Which of the following does NOT have any rectangular faces?

   A pentagonal prism

   B hexagonal pyramid

   C rectangular prism

   D square pyramid

Spiral Review

3. Without measuring, how can you determine whether two sides of a polygon are congruent? (Lesson 11.4)

   A If the two sides look congruent, you can assume they are congruent.

   B Cut out the polygon and fold the two sides onto each other. If the sides match up, you can assume they are congruent.

   C Cut out the polygon and fold two of the angles onto each other. If the angles match up, you can assume the sides are also congruent.

   D It is not possible to determine whether two sides of a polygon are congruent without measuring.

4. James has $4\frac{3}{4}$ feet of rope. He plans to cut off $1\frac{1}{2}$ feet from the rope. How much rope will be left? (Lesson 6.6)

   A $\frac{1}{4}$ foot

   B 3 feet

   C $3\frac{1}{4}$ feet

   D $6\frac{1}{2}$ feet

5. Latasha made 128 ounces of punch. How many cups of punch did Latasha make? (Lesson 10.2)

   A 4 cups

   B 8 cups

   C 16 cups

   D 32 cups

6. Which of the following statements is NOT true? (Lesson 11.3)

   A Some quadrilaterals are squares.

   B All rhombuses are quadrilaterals.

   C All squares are rectangles.

   D Some trapezoids are parallelograms.

Name _____

Unit Cubes and Solid Figures

Count the number of cubes used to build each solid figure.

1.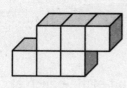

__18__ unit cubes

2.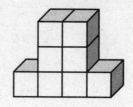

_____ unit cubes

3.

_____ unit cubes

4.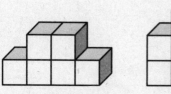

_____ unit cubes

5.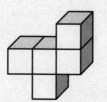

_____ unit cubes

6.

_____ unit cubes

Compare the number of unit cubes in each solid figure. Use <, >, or =.

7.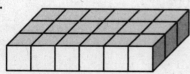

_____ unit cubes ◯ _____ unit cubes

8.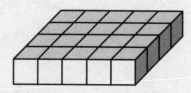

_____ unit cubes ◯ _____ unit cubes

Problem Solving REAL WORLD

9. A carton can hold 1,000 unit cubes that measure 1 inch by 1 inch by 1 inch. Describe the dimensions of the carton using unit cubes.

_____

_____

_____

10. Peter uses unit cubes to build a figure in the shape of the letter X. What is the fewest unit cubes that Peter can use to build the figure?

_____

Lesson Check

1. Cala stacked some blocks to make the figure below. How many blocks are in Cala's figure?

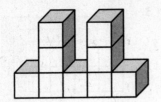

   (A) 7

   (B) 8

   (C) 9

   (D) 10

2. Quentin has 18 unit cubes. How many different rectangular prisms can he build if he uses all of the cubes?

   (A) 4

   (B) 6

   (C) 8

   (D) 18

Spiral Review

3. In what shape are the lateral faces of a pyramid? **(Lesson 11.5)**

   (A) triangle

   (B) square

   (C) rectangle

   (D) hexagon

4. The Arnold family arrived at the beach at 10:30 A.M. They spent $3\frac{3}{4}$ hours there. What time did they leave the beach? **(Lesson 10.7)**

   (A) 1:15 P.M.

   (B) 2:15 P.M.

   (C) 3:15 P.M.

   (D) 3:45 P.M.

5. Which of the following is always true about a parallelogram? **(Lesson 11.3)**

   (A) All sides are congruent.

   (B) All angles are congruent.

   (C) It has 4 right angles.

   (D) Opposite sides are congruent.

6. The tire on Frank's bike moves 75 inches in one rotation. How many rotations will the tire have made after Frank rides 50 feet?

   (Lesson 10.4)

   (A) 2

   (B) 8

   (C) 12

   (D) 24

Understand Volume

Use the unit given. Find the volume.

1.

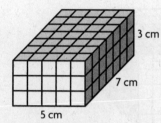

3 cm
7 cm
5 cm
Each cube = 1 cu cm

Volume = __105__ cu __cm__

2.

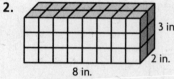

3 in.
2 in.
8 in.
Each cube = 1 cu in.

Volume = _____ cu _____

3.

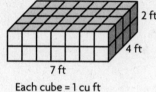

2 ft
4 ft
7 ft
Each cube = 1 cu ft

Volume = _____ cu _____

4.

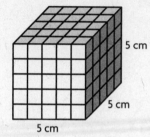

5 cm
5 cm
5 cm
Each cube = 1 cu cm

Volume = _____ cu _____

5. Compare the volumes. Write <, >, or =.

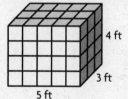

4 ft
3 ft
5 ft
Each cube = 1 cu ft

_____ cu ft ◯ _____ cu ft

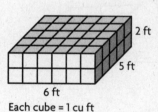

2 ft
5 ft
6 ft
Each cube = 1 cu ft

Problem Solving 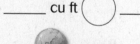 REAL WORLD

6. A manufacturer ships its product in boxes with edges of 4 inches. If 12 boxes are put in a carton and completely fill the carton, what is the volume of the carton?

_____

7. Matt and Mindy each built a rectangular prism that has a length of 5 units, a width of 2 units, and a height of 4 units. Matt used cubes that are 1 cm on each side. Mindy used cubes that are 1 in. on each side. What is the volume of each prism?

_____

_____

Lesson Check

1. Elena packed 48 cubes into this box. Each cube has edges that are 1 centimeter. How many layers of cubes did Elena make?

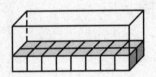

- (A) 2
- (B) 3
- (C) 4
- (D) 8

2. What is the volume of the rectangular prism?

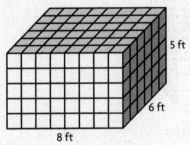

5 ft

6 ft

8 ft

Each cube = 4 cu ft

- (A) 40 cubic inches
- (B) 40 cubic feet
- (C) 240 cubic inches
- (D) 240 cubic feet

Spiral Review

3. Juan made a design with polygons. Which polygon in Juan's design is a pentagon?

(Lesson 11.1)

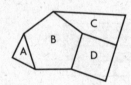

- (A) Figure A
- (C) Figure C
- (B) Figure B
- (D) Figure D

4. Which ordered pair describes the location of point X? (Lesson 9.2)

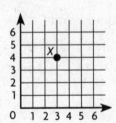

- (A) (3, 4)
- (C) (4, 4)
- (B) (4, 3)
- (D) (3, 3)

5. What is the least number of acute angles that a triangle can have? (Lesson 11.2)

- (A) 0
- (B) 1
- (C) 2
- (D) 3

6. Karen bought 3 pounds of cheese to serve at a picnic. How many ounces of cheese did Karen buy? (Lesson 10.3)

- (A) 24 ounces
- (B) 32 ounces
- (C) 36 ounces
- (D) 48 ounces

Estimate Volume

Estimate the volume.

1. Volume of package of paper: 200 cu in.

Think: Each package of paper has a volume

of 200 cu in. There are __8__ packages of

paper in the larger box. So, the volume of the

large box is about __8__ × 200, or __1,600__

cubic inches.

Volume of large box: __1,600 cu in.__

2. Volume of rice box: 500 cu cm

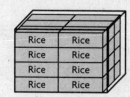

Volume of large box: _____

3. Volume of tea box: 40 cu in.

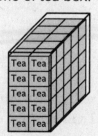

Volume of large box: _____

4. Volume of DVD case: 20 cu in.

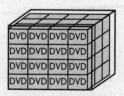

Volume of large box: _____

Problem Solving

5. Theo fills a large box with boxes of staples. The volume of each box of staples is 120 cu cm. Estimate the volume of the large box.

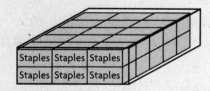

_____

6. Lisa uses pudding boxes to estimate the volume of the box below. The volume of each pudding box is 150 cu in. Estimate the volume of the large box.

_____

Lesson Check

1. Melanie packs boxes of envelopes into a larger box. The volume of each box of envelopes is 1,200 cubic centimeters. Which is the best estimate for the volume of the large box?

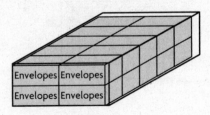

   Ⓐ 2,400 cu cm

   Ⓑ 12,000 cu cm

   Ⓒ 20,000 cu cm

   Ⓓ 24,000 cu cm

2. Calvin packs boxes of greeting cards into a larger box. The volume of each box of greeting cards is 90 cubic inches. Which is the best estimate for the volume of the large box?

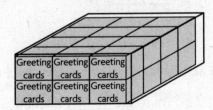

   Ⓐ 216 cu in.

   Ⓑ 240 cu in.

   Ⓒ 2,160 cu in.

   Ⓓ 2,400 cu in.

Spiral Review

3. Rosa has 16 unit cubes. How many different rectangular prisms can she build with the cubes? (Lesson 11.6)

   Ⓐ 2

   Ⓑ 3

   Ⓒ 4

   Ⓓ 8

4. Each cube represents 1 cubic inch. What is the volume of the prism? (Lesson 11.7)

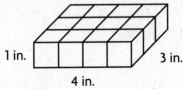

   1 in.   3 in.

   4 in.

   Ⓐ 1 cu in.

   Ⓑ 4 cu in.

   Ⓒ 8 cu in.

   Ⓓ 12 cu in.

5. A certain aquarium holds 20 gallons of water. How many quarts of water does the aquarium hold? (Lesson 10.2)

   Ⓐ 160 quarts   Ⓒ 10 quarts

   Ⓑ 80 quarts   Ⓓ 5 quarts

6. Monique ran in a 5-kilometer race. How many meters did Monique run? (Lesson 10.5)

   Ⓐ 50 meters   Ⓒ 5,000 meters

   Ⓑ 500 meters   Ⓓ 50,000 meters

Volume of Rectangular Prisms

Find the volume.

1.

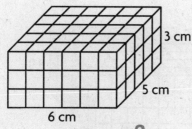

Volume: **90 cm³**

2.

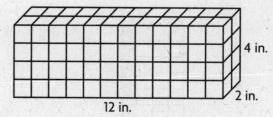

Volume: _____

3.

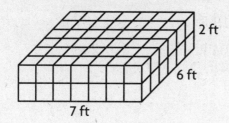

Volume: _____

4.

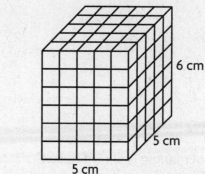

Volume: _____

5.

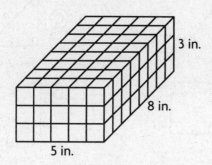

Volume: _____

6.

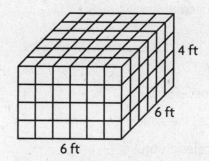

Volume: _____

Problem Solving REAL WORLD

7. Aaron keeps his baseball cards in a cardboard box that is 12 inches long, 8 inches wide, and 3 inches high. What is the volume of this box?

_____

8. Amanda's jewelry box is in the shape of a cube that has 6-inch edges. What is the volume of Amanda's jewelry box?

_____

Lesson Check

1. Laini uses 1-inch cubes to build the box shown below. What is the volume of the box?

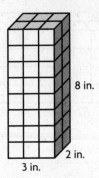

8 in.

2 in.

3 in.

(A) 13 in.³   (C) 46 in.³

(B) 16 in.³   (D) 48 in.³

2. Mason stacked 1-foot cube-shaped boxes in a warehouse. What is the volume of the stack of boxes?

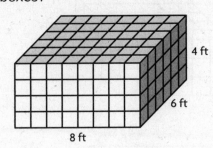

4 ft

6 ft

8 ft

(A) 19 ft³   (C) 192 ft³

(B) 104 ft³   (D) 208 ft³

Spiral Review

3. What type of triangle is shown below? **(Lesson 11.2)**

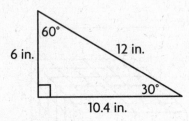

60°

6 in.

12 in.

30°

10.4 in.

(A) isosceles right

(B) scalene acute

(C) scalene obtuse

(D) scalene right

4. Which figure is a quadrilateral that has opposite sides that are congruent and parallel? **(Lesson 11.4)**

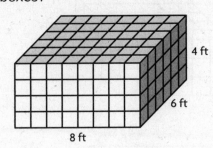

5. Suzanne is 64 inches tall. Which of the following is equal to 64 inches? **(Lesson 10.1)**

(A) 4 ft   (C) 6 ft

(B) 5 ft 4 in.   (D) 6 ft 4 in.

6. Trevor bought 8 gallons of paint to paint his house. He used all but 1 quart. How many quarts of paint did Trevor use? **(Lesson 10.4)**

(A) 7 quarts   (C) 31 quarts

(B) 15 quarts   (D) 47 quarts

Name _____

Apply Volume Formulas

Find the volume.

1.

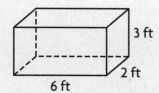

3 ft
2 ft
6 ft

$V = \underline{\quad l \quad} \times \underline{\quad w \quad} \times \underline{\quad h \quad}$

$V = \underline{\quad 6 \quad} \times \underline{\quad 2 \quad} \times \underline{\quad 3 \quad}$

$V = \underline{\quad 36 \text{ ft}^3 \quad}$

2.

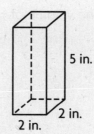

5 in.
2 in.
2 in.

$V = \underline{\hspace{3cm}}$

3.

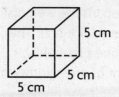

5 cm
5 cm
5 cm

$V = \underline{\hspace{3cm}}$

4.

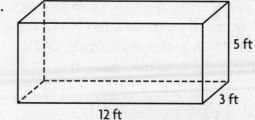

5 ft
3 ft
12 ft

$V = \underline{\hspace{3cm}}$

5.

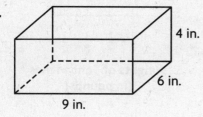

4 in.
6 in.
9 in.

$V = \underline{\hspace{3cm}}$

6.

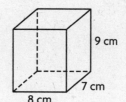

9 cm
7 cm
8 cm

$V = \underline{\hspace{3cm}}$

Problem Solving REAL WORLD

7. A construction company is digging a hole for a swimming pool. The hole will be 12 yards long, 7 yards wide, and 3 yards deep. How many cubic yards of dirt will the company need to remove?

_____

8. Amy rents a storage room that is 15 feet long, 5 feet wide, and 8 feet. What is the volume of the storage room?

_____

Lesson Check

1. Sayeed is buying a crate for his puppy. The crate is 20 inches long, 13 inches wide, and 16 inches high. What is the volume of the crate?

   (A) 208 in.³

   (B) 260 in.³

   (C) 2,600 in.³

   (D) 4,160 in.³

2. Brittany has a gift box in the shape of a cube. Each side of the box measures 15 centimeters. What is volume of the gift box?

   (A) 255 cm³

   (B) 1,350 cm³

   (C) 3,375 cm³

   (D) 3,475 cm³

Spiral Review

3. Max packs cereal boxes into a larger box. The volume of each cereal box is 175 cubic inches. Which is the best estimate for the volume of the large box? **(Lesson 11.8)**

   (A) 210 cu in.

   (B) 420 cu in.

   (C) 2,100 cu in.

   (D) 4,200 cu in.

4. In health class, students record the weights of the sandwiches they have for lunch. The weights are shown in the line plot below. What is the average weight of one sandwich? **(Lesson 9.1)**

   (A) $\frac{1}{8}$ lb

   (B) $\frac{1}{4}$ lb

   (C) $\frac{3}{10}$ lb

   (D) $\frac{3}{8}$ lb

   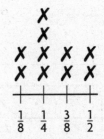

   Weights of Sandwiches (in pounds)

5. Chloe has 20 unit cubes. How many different rectangular prisms can she build with the cubes? **(Lesson 11.6)**

   (A) 4

   (B) 5

   (C) 10

   (D) 20

6. Darnell went to the movies with his friends. The movie started at 2:35 P.M. and lasted 1 hour 45 minutes. What time did the movie end? **(Lesson 10.7)**

   (A) 3:50 P.M.

   (B) 4:20 P.M.

   (C) 4:50 P.M.

   (D) 5:20 P.M.

Name _____

Problem Solving • Compare Volumes

Make a table to help you solve each problem.

1. Amita wants to make a mold for a candle. She wants the shape of the candle to be a rectangular prism with a volume of exactly 28 cubic centimeters. She wants the sides to be in whole centimeters. How many different molds can she make?

   10 molds

2. Amita decides that she wants the molds to have a square base. How many of the possible molds can she use?

   _____

3. Raymond wants to make a box that has a volume of 360 cubic inches. He wants the height to be 10 inches and the other two dimensions to be whole numbers of inches. How many different-sized boxes can he make?

   _____

4. Jeff put a small box that is 12 inches long, 8 inches wide, and 4 inches tall inside a box that is 20 inches long, 15 inches wide, and 9 inches high. How much space is left in the larger box?

   _____

5. Mrs. Nelson has a rectangular flower box that is 5 feet long and 2 feet tall. She wants the width of the box to be no more than 5 feet. If the width is a whole number, what are the possible volumes for the flower box?

   _____

6. Sophina bought 3 yards of trim to put around a rectangular scarf. She wants the width of the scarf to be a whole number that is at least 6 inches and at most 12 inches. If she uses all the trim, what are the possible dimensions of her scarf? Write your answers in inches.

   _____

Lesson Check

1. Corey bought a container shaped like a rectangular prism to hold his photo collection. If the container has a volume of 480 cubic inches, which of the following could be its dimensions in inches?

   Ⓐ 3 in. by 8 in. by 10 in.

   Ⓑ 6 in. by 4 in. by 12 in.

   Ⓒ 6 in. by 8 in. by 10 in.

   Ⓓ 8 in. by 4 in. by 8 in.

2. Aleka has a box for keepsakes that has a volume of 576 cubic inches. The length of the box is 12 inches and the width is 8 inches. What is the height of the box?

   Ⓐ 6 inches

   Ⓑ 20 inches

   Ⓒ 48 inches

   Ⓓ 72 inches

Spiral Review

3. A movie is 2 hours and 28 minutes long. It starts at 7:50 P.M. At what time will the movie end? (Lesson 10.7)

   Ⓐ 9:18 P.M.

   Ⓑ 9:78 P.M.

   Ⓒ 10:08 P.M.

   Ⓓ 10:18 P.M.

4. Which of the following does NOT have any rectangular faces? (Lesson 11.5)

   Ⓐ hexagonal prism

   Ⓑ pentagonal pyramid

   Ⓒ rectangular prism

   Ⓓ square pyramid

5. An aquarium is in the shape of a rectangular prism. Its length is 24 inches, its width is 12 inches, and its height is 14 inches. How much water can the aquarium hold? (Lesson 11.10)

   Ⓐ 168 cubic inches

   Ⓑ 288 cubic inches

   Ⓒ 336 cubic inches

   Ⓓ 4,032 cubic inches

6. What is the volume of the rectangular prism shown? (Lesson 11.9)

   Ⓐ 11 m³

   Ⓑ 18 m³

   Ⓒ 36 m³

   Ⓓ 360 m³

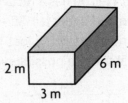

2 m   6 m
3 m

Find Volume of Composed Figures

Find the volume of the composite figure.

1.

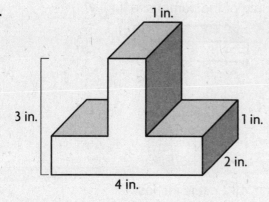

1 in.

3 in.

1 in.

2 in.

4 in.

V = _____

2.

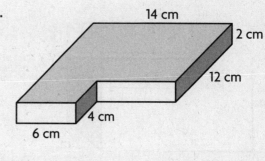

14 cm

2 cm

12 cm

4 cm

6 cm

V = _____

3.

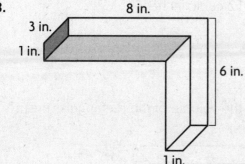

8 in.

3 in.

1 in.

6 in.

1 in.

V = _____

4.

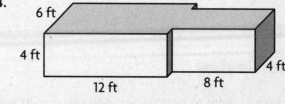

6 ft

4 ft

4 ft

12 ft

8 ft

V = _____

Problem Solving REAL WORLD

5. As part of her shop class, Jules made the figure below out of pieces of wood. How much space does the figure she made take up?

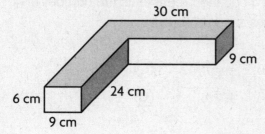

30 cm

9 cm

24 cm

6 cm

9 cm

_____

6. What is the volume of the composite figure below?

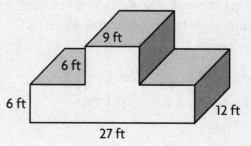

9 ft

6 ft

6 ft

12 ft

27 ft

_____

Lesson Check

1. Which expression represents the volume of the composite figure?

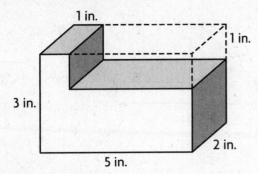

1 in.
1 in.
3 in.
2 in.
5 in.

(A) $(5 \times 2) - (3 \times 1)$

(B) $5 \times 2 \times 3$

(C) $(5 \times 2 \times 3) - (4 \times 2 \times 1)$

(D) $4 \times 2 \times 1$

2. Suppose you take the small prism and stack it on top of the larger prism. What will be the volume of the composite figure?

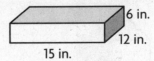

6 in.
12 in.
15 in.

6 in.
12 in.
6 in.

(A) 432 cubic inches

(B) 648 cubic inches

(C) 1,080 cubic inches

(D) 1,512 cubic inches

Spiral Review

3. Jesse wants to build a wooden chest with a volume of 8,100 cubic inches. The length will be 30 inches and the width will be 15 inches. How tall will Jesse's chest be? (Lesson 11.11)

(A) 18 in.

(B) 30 in.

(C) 270 in.

(D) 540 in.

4. What is the volume of the rectangular prism?
(Lesson 11.9)

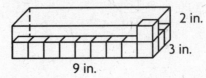

2 in.
3 in.
9 in.

(A) 14 in.^{3} (C) 45 in.^{3}

(B) 27 in.^{3} (D) 54 in.^{3}

5. Adrian's recipe for cranberry relish calls for $1\frac{3}{4}$ cups of sugar. He wants to use $\frac{1}{2}$ that amount. How much sugar should he use?
(Lesson 7.9)

(A) $1\frac{1}{4}$ cups   (C) $\frac{7}{8}$ cup

(B) $1\frac{1}{6}$ cups   (D) $\frac{1}{2}$ cup

6. Joanna has a board that is 6 feet long. She cuts it into pieces that are each $\frac{1}{4}$ foot long. Which equation represents the number of pieces she cut? (Lesson 8.5)

(A) $6 \div \frac{1}{4} = n$ (C) $\frac{1}{4} \div 6 = n$

(B) $6 \div 4 = n$ (D) $\frac{1}{4} \div \frac{1}{6} = n$

Name _____

Chapter 11 Extra Practice

Lesson 11.1

Name each polygon. Then tell whether it is a *regular* polygon or
not a regular polygon.

1.

2.

_____   _____

Lesson 11.2

Classify each triangle. Write *isosceles*, *scalene*, or *equilateral*. Then
write *acute*, *obtuse*, or *right*.

1.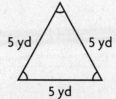

2.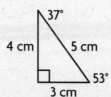

3.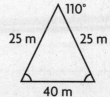

_____   _____   _____

_____   _____   _____

Lesson 11.3

Classify the quadrilateral in as many ways as possible. Write
quadrilateral, parallelogram, rectangle, rhombus, square, or *trapezoid.*

1.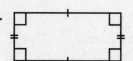

2.

_____   _____

_____   _____

_____   _____

Lesson 11.4

1. Sasha has a triangle with vertices A, B, and C. The triangle has three congruent angles. She wants to show that triangle ABC has three congruent sides, but she does not have a ruler to measure the lengths of the sides. How can she show that the triangle has three congruent sides?

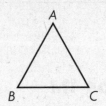

_____

_____

_____

_____

Lesson 11.5

Classify the solid figure. Write *prism, pyramid, cone, cylinder,* **or** *sphere.*

1.

_____

2.

_____

Lessons 11.6 - 11.10

Find the volume.

1.

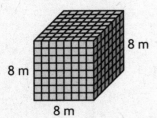

8 m

8 m

8 m

Volume = _____

2.

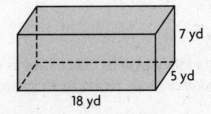

7 yd

5 yd

18 yd

Volume = _____

Lesson 11.11

Solve.

1. One aquarium is 12 inches long, another is 15 inches long, and a third is 18 inches long. They are all 18 inches deep and 12 inches wide. Which aquarium can hold exactly 3,240 cubic inches of water?

_____

Name _____

Compare Fractions and Decimals

Essential Question How can you compare decimals, fractions, and mixed numbers on a number line?

🔓 UNLOCK the Problem   REAL WORLD

The Tech Club compared the weights of three cell phones. Estéban's phone weighed 4.7 ounces. Jill's phone weighed $4\frac{3}{5}$ ounces. Mona's phone weighed 4.35 ounces. Who has the phone with the lightest weight?

You can use a number line to compare fractions and decimals.

Remember: Greater values on a number line lie farther to the right.

 Compare the values on a number line.

STEP 1 Locate some benchmarks.

• Benchmark decimals: 4, 4.25, 4.5, 4.75, 5…

• Benchmark mixed numbers: 4, $4\frac{1}{4}$, $4\frac{1}{2}$, $4\frac{3}{4}$, 5…

- How can you identify the number with the least value?

_____

_____

_____

_____

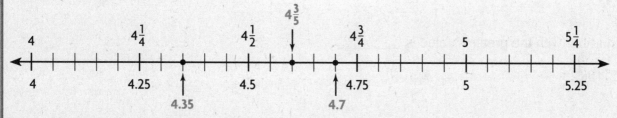

STEP 2 Mark the weight of each cell phone on the number line.

• Find the location of 4.7, $4\frac{3}{5}$, and 4.35.

Since $4.35 < 4\frac{3}{5} < 4.7$, Mona's phone is lightest.

Try This! Compare $\frac{1}{5}$, $\frac{5}{8}$, and 0.2. Which number has the greatest value?

• Mark each value on a number line.

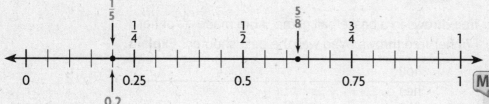

The greatest number is _____. **Explain** how you decided.

Math Talk **Explain** how you can tell that $\frac{1}{5}$ and 0.2 are equal.

Share and Show

For 1–2, identify the points on the number line.
Then write the greater number.

1. point A as a decimal

_____

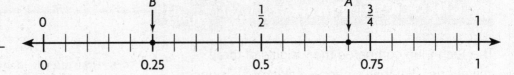

2. point B as a fraction

_____   _____ is greater.

Locate each number on a number line.
Then complete the sentence.

3. 0.55, $\frac{2}{5}$, 0.46

The number with the greatest value is _____.

On Your Own

Locate each number on a number line. Then complete the sentence.

4. 0.4, $\frac{3}{4}$, 0.15

The number with the greatest value is _____.

5. $2\frac{2}{3}$, 2.45, $2\frac{2}{5}$

The number with the least value is _____.

6 3.95, $3\frac{5}{6}$, $3\frac{4}{5}$

The number with the greatest value is _____.

Problem Solving REAL WORLD

7. Hannah made 0.7 of her free throws in a basketball game. Abra made $\frac{9}{10}$ of her free throws. Dena made $\frac{3}{4}$ of her free throws. Who was the best shooter? **Explain.**

_____

_____

Order Fractions and Decimals

Essential Question How can you order decimals, fractions, and mixed numbers on a number line?

🔓 UNLOCK the Problem REAL WORLD

In tennis, Jocelyn's serve takes 0.97 of a second to reach her opponent. Dave's serve takes $\frac{4}{5}$ of a second. Monica's serve takes 0.85 of a second. Order the three serves from shortest to longest time.

- You want to order the times from shortest to longest. Should you read the numbers on the number line left to right or right to left?

  _____

  _____

  _____

🔑 Order the fractions and decimals on the number line.

STEP 1 Locate the benchmarks on the number line.

- Benchmark decimals: 0, 0.25, 0.5, 0.75, 1.
- Benchmark fractions: 0, $\frac{1}{4}$, $\frac{1}{2}$, $\frac{3}{4}$, 1.

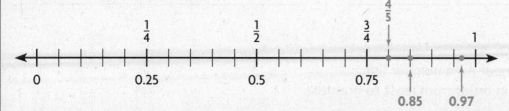

STEP 2 Locate 0.97, $\frac{4}{5}$, and 0.85 on the number line.

STEP 3 Order the fractions and decimals.

Remember: The point farthest to the left is the least value.

So, the times in order from shortest to longest are: ___$\frac{4}{5}$___, __0.85__, __0.97__ .

Try This! Order 6.03, $5\frac{9}{10}$, $5\frac{3}{4}$, and 6.2 from greatest to least.

- Locate each fraction and decimal on the number line. Use benchmarks to help you locate each.

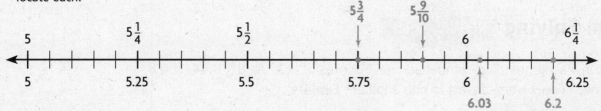

From the greatest to least: _____, _____, _____, _____

Math Talk How does the number line help you order numbers from greatest to least?

Share and Show

Locate each number on the number line.
Then write the numbers in order from least to greatest.

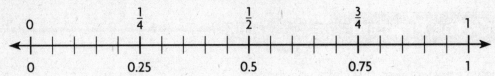

1. $\frac{3}{5}$, 0.54, 0.35

_____

For 2–3, locate each set of numbers on a number line.
Then write the numbers in order from greatest to least.

2. 1.16, $1\frac{1}{4}$, 1.37, $1\frac{1}{10}$

_____

3. $\frac{5}{8}$, 0.5, $\frac{2}{5}$, 0.78

_____

On Your Own

For 4–5, locate each number on a number line.
Then write the numbers in order from least to greatest.

4. 0.6, $\frac{1}{2}$, $\frac{2}{3}$, 0.39

_____

5. $7\frac{1}{4}$, 7.4, $7\frac{3}{4}$, 7.77

_____

For 6–7, locate each number on a number line.
Then write the numbers in order from greatest to least.

6. $\frac{3}{10}$, 0.222, $\frac{3}{5}$, 0.53

_____

7. 2.96, $3\frac{1}{5}$, 3.48, $3\frac{1}{4}$

_____

Problem Solving REAL WORLD

8. Judges in a skateboarding competition gave scores of 8.2, $8\frac{1}{3}$, $8\frac{4}{5}$, 8.44, and $8\frac{1}{5}$.
   Which two scores were closest to one another? **Explain.**

_____

_____

Name _____

Factor Trees

Essential Question How can you factor numbers using a factor tree?

 UNLOCK the Problem REAL WORLD

Mr. Shu gives this puzzle to his math students.

"Write 24 as a product of factors that are prime. Remember that a prime number must be greater than 1 and can have only 1 and itself as factors."

You can use a diagram called a **factor tree** to find the factors of a number.

- Give an example of a number greater than 1 that has only 1 and itself as factors.

_____

Use a factor tree to find the prime number factors that have a product of 24.

STEP 1	STEP 2	STEP 3	STEP 4
Write the number to be factored at the top of the factor tree.	Write it as a product of any two factors. **Think:** $4 \times 6 = 24$	Write each factor as the product of two factors. **Think:** $2 \times 2 = 4$ and $2 \times 3 = 6$	Continue until each factor is a prime number. **Think:** $2 \times 1 = 2$ and $3 \times 1 = 3$ Write the factors that are prime numbers from least to greatest.

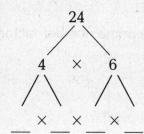

__ × __ × __ × __

So, 24 = _____ .

Try This! Make a different factor tree for 24.

- Is the product of factors the same as in the Example? **Explain.**

_____

_____

_____

Math Talk **Explain** how you can use factored numbers to find common factors.

Share and Show

1. Use a factor tree to find the prime number factors that have a product of 210.

   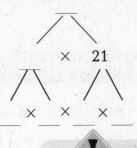

   • Write 210 as a product of any two factors.

   _____ = _____ × 21

   • Write each factor as the product of factors.

   10 = _____ × _____   21 = _____ × _____

   Now each factor has only _____ and itself as factors.

   So, 210 = _____ × _____ × _____ × _____ .

> **! ERROR Alert**
>
> Remember to continue to factor a number if it has factors other than 1 and itself.

Use a factor tree to find the prime number factors.

2. 8

3. 45

4. 350

_____ | _____ | _____

On Your Own

Use a factor tree to find the prime number factors.

5. 36

6. 72

7. 540

_____ | _____ | _____

Problem Solving REAL WORLD

Mr. Shu gave these problems to his math students. Solve.

8. Write 500 as a product of prime number factors. Each factor must be greater than 1 and can have only 1 and itself as factors.

9. Find a number that has four identical even factors. Each factor must be greater than 1 and can have only 1 and itself as factors.

_____

_____

Name _____

Model Percent

Essential Question How can you express real world quantities as percents and use them to solve problems?

🔑 UNLOCK the Problem

Percent means "per hundred" or "out of 100." So, when you find percent you are finding a part of 100. Sixty percent, for example, means 60 out of 100. You can write percents using the percent symbol, %. So, 60 percent is written as 60%.

- What number is always compared in a percent?

_____

🔒 **Example 1** Name the percent that is shaded.

- 5 columns: $5 \times 10 = 50$.
- 3 squares: $3 \times 1 = 3$
- Total: $50 + 3 = 53$ out of 100, or 53 percent is shaded.

🔒 **Example 2** Name the percent that is not shaded.

- 4 columns: $4 \times 10 = 40$.
- 7 squares: $7 \times 1 = 7$
- Total: $40 + 7 = 47$ out of 100, or 47 percent is not shaded.

Try This! Use the number line. Tell what these percents mean:
0 percent, 50 percent, 100 percent.

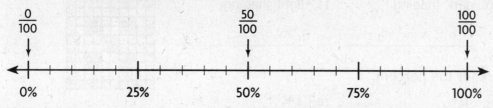

A. 0 percent means _____ out of 100, or none of the total.

B. 50 percent means _____ out of 100, or half of the total.

C. 100 percent means _____ out of 100, or all of the total.

Math Talk Which benchmark is 33% closest to? **Explain** how you know.

Share and Show

Use the diagram to write the percent.

1. How many whole columns and single squares are shaded?

   _____

2. What percent is shaded?

   _____

3. What percent is unshaded?

   _____

Shade the grid to show the percent.

4. 20 percent

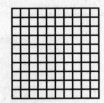

5. 86 percent

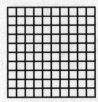

On Your Own

Use the diagram to write the percent.

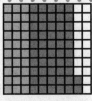

6. light shading

   _____

7. dark shading

   _____

8. not shaded

   _____

9. not shaded

   _____

10. dark shading

    _____

11. light shading

    _____

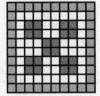

Write the closest benchmark for the percent.

12. 48%

    _____

13. 94%

    _____

14. 4%

    _____

Problem Solving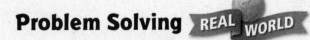

15. In an election between Warren and Jorge, Warren declared victory because he received 58 percent of the vote. Is he correct? **Explain**.

    _____

    _____

Name _____

Relate Decimals and Percents

Essential Question How can you express decimals as percents and percents as decimals?

 **UNLOCK the Problem** REAL WORLD

Decimals and percents are two ways of expressing the same number. You can write a percent as a decimal. You can also write a decimal as a percent.

- In percent, the "whole" is 100. What is the "whole" in decimal form?

  _____

Example 1 Model 0.42. Write 0.42 as a percent.

STEP 1 Write the decimal as a ratio.

0.42 = 42 hundredths = 42 out of 100.

_____

STEP 2 Make a model that shows 42 out of 100.

_____

STEP 3 Use the model to write a percent.

42 shaded squares = __42__ percent, or __42__%

Example 2 Model 19 percent. Write 19% as a decimal.

STEP 1 Write the percent as a fraction.

$19\% = \frac{19}{100}$

_____

STEP 2 Make a model that shows 19 out of 100.

_____

STEP 3 Use the model to write a decimal.

19 shaded squares out of 100 squares = _____

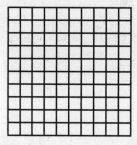

Math Talk Suppose a store is having a 50% off sale. What does this mean?

Share and Show

Use the model. Complete each statement.

1a. $0.68 = $ _____ out of 100

1b. How many squares are shaded?

_____

1c. What percent is shaded?

_____

Write the percents as decimals.

2. 47 percent

_____

3. 11 percent

_____

On Your Own

Write the decimals as percents.

4. 0.20

_____

5. 0.39

_____

6. 0.44

_____

7. 0.93

_____

8. 0.07

_____

9. 0.7

_____

10. 0.06

_____

11. 0.6

_____

Write the percents as decimals.

12. 12 percent

_____

13. 31%

_____

14. 99 percent

_____

15. 13 percent

_____

16. 4 percent

_____

17. 14 percent

_____

18. 90 percent

_____

19. 9%

_____

Problem Solving REAL WORLD

20. In basketball, Linda made 0.56 of her shots. What percent of her shots did Linda miss?

_____

Name _____

Fractions, Decimals, and Percents

Essential Question How can you convert between fractions, decimals, and percents?

UNLOCK the Problem REAL WORLD

Every percent and decimal number can also be written as a fraction. All fractions can be written as decimals and percents. For example, $\frac{2}{5}$ of the songs in Bonnie's music collection are country songs. What percent of her song collection is country?

 Write the percent that is equivalent to $\frac{2}{5}$.

STEP 1   Set up the equivalent fraction with a denominator of 100.

$$\frac{2 \times ?}{5 \times ?} = \frac{}{100}$$

STEP 2   Ask: By what factor can you multiply the denominator to get 100?

$$\frac{2 \times ?}{5 \times 20} = \frac{}{100} \longleftarrow \text{multiply the denominator by 20}$$

STEP 3   Multiply the numerator by the same factor, 20.

$$\frac{2 \times 20}{5 \times 20} = \frac{40}{100}$$

STEP 4   Write the fraction as a percent.

$$\frac{40}{100} = \underline{40} \text{ percent}$$

So, $\frac{2}{5}$ equals $\underline{40}$ percent.

More Examples

A. Write $\frac{8}{25}$ as a decimal.

STEP 1   Write an equivalent fraction with a denominator of 100.

$$\frac{8 \times 4}{25 \times 4} = \frac{32}{100} \longleftarrow \text{multiply denominator and numerator by 4}$$

STEP 2   Write the fraction as a decimal.

$$\frac{32}{100} = 0.32$$

B. Write 90 percent as a fraction in simplest form.

STEP 1   Write 90% as a fraction.

$$90\% = \frac{90}{100}$$

STEP 2   Simplify.

$$90\% = \frac{90 \div 10}{100 \div 10} = \frac{9}{10}$$

Math Talk   How are 9% and 90% alike when written as decimals? How are they different?

Share and Show

Complete the steps to write $\frac{7}{20}$ as a percent.

1. By what factor should you multiply the

   denominator and numerator? _____

   $$\frac{7 \times ?}{20 \times ?} = \frac{?}{100}$$

2. For $\frac{7}{20}$, what is an equivalent fraction with a denominator of 100?

3. What percent is equivalent to $\frac{7}{20}$?

_____   _____

Write a decimal, a percent, or a simplified fraction.

4. $\frac{1}{4}$ as a decimal

5. $\frac{3}{10}$ as a percent

6. 80% as a fraction

_____   _____   _____

On Your Own

Write a decimal, a percent, or a simplified fraction.

7. $\frac{1}{2}$ as a percent

8. $\frac{9}{10}$ as a decimal

9. $\frac{11}{20}$ as a percent

10. 75% as a fraction

_____   _____   _____   _____

11. $\frac{3}{5}$ as a percent

12. $\frac{9}{25}$ as a decimal

13. $\frac{29}{50}$ as a percent

14. $\frac{1}{20}$ as a percent

_____   _____   _____   _____

15. 4% as fraction

16. $\frac{4}{5}$ as a percent

17. $\frac{24}{25}$ as a decimal

18. $\frac{41}{50}$ as a percent

_____   _____   _____   _____

Problem Solving REAL WORLD

19. Whitney has finished $\frac{9}{20}$ of her book. What percent of the book does Whitney still need to read?

20. Roger has completed $\frac{4}{25}$ of his math homework. What percent of his math homework does he still need to do?

_____   _____

Name _____

 Checkpoint

Concepts and Skills

Locate each number on the number line. Then complete the sentence. (pp. P245–P246)

1. $0.4, \frac{3}{5}, 0.35$

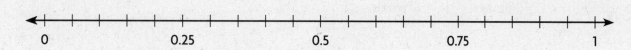

0   0.25   0.5   0.75   1

The number with the least value is _____.

Write the numbers in order from least to greatest. (pp. P247–P248)

2. $0.4, \frac{3}{5}, 0.55, \frac{1}{4}$

3. $\frac{3}{4}, 0.7, \frac{1}{2}, 0.1$

_____   _____

Use a factor tree to find the prime number factors. (pp. P249–P250)

4. 16

5. 36

6. 42

_____   _____   _____

Write a decimal, a percent, or a simplified fraction. (pp. P251–P256)

7. 0.08 as a percent   8. $\frac{3}{5}$ as a decimal   9. 80% as a fraction   10. $\frac{13}{20}$ as a percent

_____   _____   _____   _____

Problem Solving REAL WORLD

For 11–12, use the data in the table. (pp. P251–P256)

11. What percent of the apes in the Wild Country Zoo are orangutans?

_____

12. One species makes up 40% of the apes in the zoo. Which species is it?

_____

| Apes in the Wild Country Zoo ||
Species	Number
Bonobo	4
Chimpanzee	20
Gorilla	15
Orangutan	11
Total	**50**

Fill in the bubble or grid completely to show your answer.

13. Entries for the Lake Manatee Bass Fishing Contest are shown.
First place is awarded to the contestant with the heaviest fish.

Lake Manatee Bass Contest	
Contestant	Weight of fish caught
George	6.25 pounds
Mia	$6\frac{2}{5}$ pounds
Harvey	$6\frac{1}{3}$ pounds

What is the correct order from first place to third place? **(pp. P247–P248)**

(A) First: George, Second: Mia, Third: Harvey

(B) First: Mia, Second: George, Third: Harvey

(C) First: Mia, Second: Harvey, Third: George

(D) First: Harvey, Second: Mia, Third: George

14. Ric used a factor tree to write 180 as a product of factors that are prime numbers. How many factors were in Ric's product? **(pp. P249–P250)**

(A) 2

(B) 3

(C) 4

(D) 5

15. On Monday, 6% of the students at Riverside School were absent. Written as a decimal, what portion of Riverside's students attended school that day? **(pp. P253–P254)**

(A) 0.06

(B) 0.6

(C) 0.94

(D) 9

16. The Hastings family drove $\frac{12}{25}$ of the distance to Yellowstone National Park on the first day of their vacation. What percent of the distance to the park remained for them to drive? **(pp. P255–P256)**

(A) 12%   (C) 48%

(B) 13%   (D) 52%

Name _____

Divide Fractions by a Whole Number

Essential Question  How do you divide a fraction by a whole number?

 UNLOCK the Problem REAL WORLD

Four friends share $\frac{2}{3}$ of a quart of ice cream equally. What fraction of a quart of ice cream does each friend get?

- What operation will you use to solve the problem?

  _____

🔑 **Divide.** $\frac{2}{3} \div 4$

STEP 1	**STEP 2**	**STEP 3**
Let the rectangle represent 1 quart of ice cream. Divide it into thirds by drawing vertical lines. Shade 2 of the thirds.	Divide the rectangle into fourths by drawing horizontal lines. Shade $\frac{1}{4}$ of the $\frac{2}{3}$ already shaded.	The rectangle is now divided into _____ equal parts. Each part is _____ of the rectangle. Of the 12 equal parts, _____ parts are shaded twice. So, _____ of the rectangle is shaded twice.

 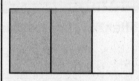 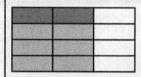

So, each friend gets _____ of a quart of ice cream.

Math Talk **Explain** why you divided the rectangle into fourths in Step 2.

Try This! **Divide.** $\frac{3}{4} \div 2$

STEP 1	**STEP 2**	**STEP 3**
Divide the rectangle into fourths. Shade 3 of the fourths.	Divide the rectangle into halves. Shade $\frac{1}{2}$ of the $\frac{3}{4}$ already shaded.	Of the 8 equal parts, _____ parts are shaded twice. So, _____ of the rectangle is shaded twice.

  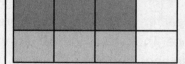

So, $\frac{3}{4} \div 2 =$ _____ .

Share and Show

Complete the model to find the quotient. Write the quotient in simplest form.

1. $\frac{5}{6} \div 2 =$ _____

Divide the rectangle into sixths.
Shade 5 of the sixths.

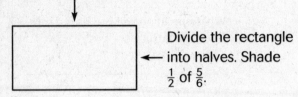

Divide the rectangle into halves. Shade $\frac{1}{2}$ of $\frac{5}{6}$.

2. $\frac{3}{4} \div 3 =$ _____

3. $\frac{2}{3} \div 3 =$ _____

4. $\frac{3}{5} \div 2 =$ _____

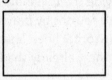

On Your Own

Complete the model to find the quotient. Write the quotient in simplest form.

5. $\frac{2}{5} \div 2 =$ _____

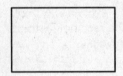

6. $\frac{5}{8} \div 3 =$ _____

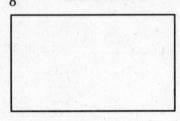

Draw a model to find the quotient. Write the quotient in simplest form.

7. $\frac{4}{9} \div 2 =$ _____

8. $\frac{4}{5} \div 3 =$ _____

Problem Solving REAL WORLD

9. Heather, Jocelyn, and Dane are each swimming one leg of a $\frac{9}{10}$-mile race. They will divide the distance equally. How far will each team member swim?

_____

Name _____

Ratios

Essential Question How can you express real world quantities as ratios?

 **UNLOCK the Problem** REAL WORLD

Max sells bouquets of roses. There are 3 yellow roses and 2 red roses. What is the ratio of yellow to red roses?

A ratio is a comparison of two numbers.

- A ratio is expressed by comparing one part to another, such as 4 feet to 20 toes, or 3 yellow roses to

  _____.

 **Activity**

Materials ■ two-color counters

Model the data.

STEP 1 Use 3 counters with the yellow side up to represent yellow roses and 2 counters with the red side up to represent red roses.

_____

STEP 2 Write the ratio of yellow to red roses.

- Ratios can be written in different ways.

  3 to 2   or   3:2   or $\frac{3}{2}$ (as a fraction)

So, the ratio of yellow roses to red roses is ___3 to 2___, ___3:2___, or ___$\frac{3}{2}$___.

In the example above, you compared a part to a part. You can also use a ratio to compare a part to a whole or a whole to a part.

Try This! Show a ratio of red counters to total counters.

STEP 1   Count to find the number of red counters. _____

_____

STEP 2   Count to find the total number of counters. _____

_____

STEP 3   Write the ratio. _____

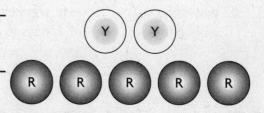

Math Talk How would the ratio change if you found the ratio of total counters to red counters?

Share and Show

Find the ratio of red counters to yellow counters.

1a. How many red counters are there?

_____

1b. How many yellow counters are there?

_____

1c. What is the ratio of red to yellow counters?

_____

Write the ratio.

2. squares to circles

_____

3. total squares to dark squares

_____

On Your Own ·

For 4–6, use the drawing to write the ratio.

4. dark to light

5. light to dark

6. light to total

_____   _____   _____

For 7–9, use the drawing to write the ratio.

7. triangles to circles

8. dark to light

9. total shapes to circles

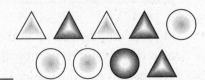

_____   _____   _____

For 10–12, write the ratio.

10. weekdays to weekend days

11. weekend days to days in a week

12. days in a week to days in January

_____   _____   _____

Problem Solving REAL WORLD

13. The ratio of length to width in Gus's driveway is 13 yards to 4 yards. What is this ratio in feet? (Hint: 3 ft = 1 yd)

_____

Equivalent Ratios

Essential Question How can you determine if two ratios are equivalent?

🔓 UNLOCK the Problem · REAL WORLD

To make brass, you can mix 2 parts zinc to 3 parts copper, a ratio of 2 to 3. If you have 12 bars of copper and use them all, how many bars of zinc do you need to make brass?

Since ratios can be written as fractions, 2 to 3 can be written as $\frac{2}{3}$. Use what you know about equivalent fractions to find equivalent ratios.

> • You know that each group of zinc to copper bars needed to make brass has a ratio of 2 to 3. How can you use this group to find an equivalent ratio?

🔑 **Use a diagram to find an equivalent ratio.**

 zinc

STEP 1 Draw bars to represent a 2 to 3 ratio of zinc to copper.

 copper

STEP 2 Add groups until you have 12 bars of copper.

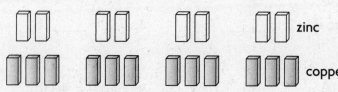

STEP 3 Count the zinc bars. Write an equivalent ratio.

There are 8 zinc bars. So, 2 to 3 is equivalent to the ratio 8 to 12.

Try This! Use equivalent ratios to find out if 6:8 is equivalent to 18:24.

STEP 1 Write the ratios as fractions.

$$6:8 = \frac{6}{8} \qquad 18:24 = \frac{18}{24}$$

STEP 2 Write the fractions in simplest form. Then compare.

$$\frac{6 \div 2}{8 \div 2} = \frac{3}{4} \qquad \frac{18 \div 6}{24 \div 6} = \frac{3}{4}$$

Both ratios equal $\frac{3}{4}$, so they are equivalent.

> **Math Talk** How does knowing how to simplify fractions help you decide whether two ratios are equivalent?

Share and Show

Are the ratios 3:5 and 12:20 equivalent?

1a. Write both ratios as fractions.

_____

1b. Are both ratios in simplest form?

_____

1c. Write both ratios in simplest form.

_____

1d. Are the ratios equivalent?

_____

Write *equivalent* or *not equivalent.*

2. 1 to 3 and 2 to 6

_____

3. 3 to 7 and 12 to 21

_____

On Your Own ·

Write the equivalent ratio.

4. 5 to 2 = _____ to 4

5. 3 to 6 = 7 to _____

6. 7:2 = _____ :6

7. 14 to 21 = _____ to 15

8. 6:10 = _____ :30

9. 8 to 9 = 40 to _____

Write *equivalent* or *not equivalent.*

10. 3:5 and 21:35

_____

11. 4 to 3 and 36 to 24

_____

12. 27:72 and 9:24

_____

Problem Solving REAL WORLD

13. Three of every 5 pizzas that Miggy's Pizza sells are cheese pizzas. Miggy's sold 80 pizzas today. How many of them would you expect were cheese?

_____

Name _____

Rates

Essential Question How can you find rates and unit rates?

🔓 UNLOCK the Problem REAL WORLD

CONNECT You know how to write ratios to compare two quantities. A **rate** is a ratio that compares two quantities that have different units of measure. A **unit rate** is a rate that has 1 unit as its second term.

Rafael is shopping at a used book and music store. A sign advertises 4 CDs for $12. What is the unit rate for the cost of 1 CD?

- What are the units of the quantities that are being compared?

  _____

- What operations can you use to write equivalent ratios?

  _____

  _____

🔑 **Write the rate in fraction form. Then find the unit rate.**

STEP 1

Write the rate in fraction form to compare dollars to CDs.

$\dfrac{\text{dollars} \longrightarrow}{\text{CDs} \longrightarrow} \dfrac{12}{\square}$

STEP 2

Divide to find an equivalent rate so that 1 is the second term.

$\dfrac{12}{4} = \dfrac{12 \div \square}{4 \div \square} = \dfrac{\square}{1}$ ← unit rate

So, the unit rate for CDs is _____ for 1 CD.

Math Talk Would it make sense to compare CDs to dollars to find a unit rate? **Explain.**

- **What if** the regular price of CDs is 5 for $20? What is the unit rate for CDs at the regular price? **Explain** how you found your answer.

  _____

  _____

  _____

Share and Show

1. Find the unit rate of speed for 120 miles in 2 hours.

$$\frac{\text{miles} \longrightarrow 120}{\text{hours} \longrightarrow \boxed{}} = \frac{\boxed{} \div 2}{2 \div \boxed{}} = \frac{\boxed{}}{\boxed{}}$$

The unit rate of speed is _____ per _____.

Find the unit rate.

2. $5.00 for 2 T-shirts

3. 200 words in 4 min

4. 150 mi on 10 gal of gas

_____

_____

_____

On Your Own

Write the rate in fraction form.

5. 90 words in 2 min

6. $1.20 for 6 goldfish

7. $0.05 per page

_____

_____

_____

Find the unit rate.

8. $208 for 4 tires

9. 300 mi per 15 gal

10. 240 people per 2 sq mi

_____

_____

_____

Problem Solving REAL WORLD

11. An ice skating rink charges $1.50 to rent ice skates for 30 minutes. What is the unit rate per hour for renting ice skates?

_____

Name _____

Distance, Rate, and Time

Essential Question How can you solve problems involving distance, rate, and time?

🔑 UNLOCK the Problem REAL WORLD

You can use the formula $d = r \times t$ to solve problems involving distance, rate, and time. In the formula, d represents distance, r represents rate, and t represents time. The rate is usually a unit rate comparing distance to time, such as miles per hour.

- What word is used in place of rate?

_____

- What are the given values?

_____

- What is the unknown value?

_____

🔑 Example 1

The winner of an automobile race drove 500 miles at an average speed of 150 miles per hour. How long did it take the winner to finish the race?

STEP 1

Write the formula.

$d = r \times t$

STEP 2

Replace d with 500 and r with 150.

$d = r \times t$

$500 = \boxed{} \times t$

STEP 3

Use what you know about inverse operations to find t.

$500 \div \boxed{} = t$

$3\frac{1}{3} = t$

So, it takes the winner _____ hours or _____ hours _____ minutes to complete the race.

🔑 Example 2

A race car driver traveled at an average speed of 120 miles per hour to finish a race in 2 hours. What was the length of the race?

STEP 1

Write the formula.

$d = r \times t$

STEP 2

Replace r with 120 and t with 2.

$d = r \times t$

$d = \boxed{} \times \boxed{}$

STEP 3

Multiply to solve for d.

$d = 120 \times 2$

$d = \boxed{}$

So, the race was _____ miles long.

Math Talk Why were different operations used in Step 3 of Examples 1 and 2?

Share and Show

1. A cyclist travels 45 miles in 3 hours. What is the cyclist's speed?

    Write the formula: $d = \boxed{} \times \boxed{}$

    Replace d with _____.

    Replace t with _____.

    The rate is _____ miles per hour.

Use the formula $d = r \times t$ to solve. Include the units in your answer.

2. A train travels at an average speed of 80 miles per hour for 5 hours. How far does the train travel?

    _____

3. A horse travels at an average speed of 12 miles per hour. How long does it take the horse to travel 60 miles?

    _____

On Your Own . . . .

Use the formula $d = r \times t$ to solve. Include the unit in your answer.

4. A hiker travels at a speed of 3 miles per hour for 3 hours. How far does the hiker travel in that time?

    _____

5. A snail travels at a speed of 2 centimeters per minute. How long does the snail take to travel 30 centimeters?

    _____

6. A boat travels 6 miles in 24 minutes. What is the average speed of the boat?

    _____

7. $d = 320$ cm

    $r =$ _____

    $t = 8$ sec

8. $d =$ _____

    $r = 50$ km per hr

    $t = 6$ hr

9. $d = 150$ ft

    $r = 20$ ft per min

    $t =$ _____

    _____

Problem Solving

10. In an experiment, Ava found that it took a ball 5 seconds to roll down an 80-foot ramp. What is the average speed of the ball?

    _____

11. Jason's family is driving 1,375 miles to Grand Canyon National Park. They plan to drive at an average speed of 55 miles per hour. How long will they be driving to reach the park?

    _____

Name _____

Concepts and Skills

Draw a model to find the quotient. Write the quotient in simplest form. (pp. P259–P260)

1. $\frac{3}{4} \div 3$

2. $\frac{2}{3} \div 5$

3. $\frac{3}{7} \div 2$

_____   _____   _____

For 4–6, use the drawing to write the ratio. (pp. P261–P262)

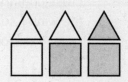

4. squares to triangles

5. total to dark

6. triangles to total

_____   _____   _____

Write the equivalent ratio. (pp. P263–P264)

7. 8 to 3 = ____ to 12

8. 2 to 6 = 4 to ____

9. 11:4 = ____ :16

Find the unit rate. (pp. P265–P266)

10. 45 visitors with 5 tour guides

11. 450 mi on 15 gal of gas

12. $56 in 8 hr

_____   _____   _____

Use the formula $d = r \times t$ to solve the problem. Include the units in your answer. (pp. P267–P268)

13. $d =$ _____

    $r = 40$ km per hr

    $t = 3$ hr

14. $d = 90$ ft

    $r = 10$ ft per sec

    $t =$ _____

15. $d = 300$ mi

    $r =$ _____

    $t = 4$ hr

Problem Solving REAL WORLD

Use the table for 16–17. (pp. P265–P268)

16. Fuel efficiency can be written as a rate comparing the distance driven to the gallons of gas used. What is the fuel efficiency of Car A written as a unit rate?

    _____

17. During the test, Car B was driven at the speed of 48 miles per hour. How long did the test take?

    _____

Fuel Test Results		
Car	Distance (in mi)	Gas (in gal)
A	308	14
B	288	12

Fill in the bubble completely to show your answer.

18. To make fruit punch for a party, Alison used 3 quarts of pineapple juice and 2 gallons of orange juice. There are 4 quarts in a gallon. What is the ratio of pineapple to orange juice in quarts? **(pp. P261–P262)**

   Ⓐ 3 to 2

   Ⓑ 3 to 5

   Ⓒ 3 to 8

   Ⓓ 8 to 3

19. Three out of every 10 pairs of skis sold by Snow Sports are cross-country skis. Snow Sports sold 450 pairs of skis during the winter season. How many of the skis were likely to have been cross-country skis? **(pp. P263–P264)**

   Ⓐ 443

   Ⓑ 135

   Ⓒ 45

   Ⓓ 30

20. At Greentree Elementary School, there are 72 fifth graders in 3 classrooms. What unit rate describes this situation? **(pp. P265–P266)**

   Ⓐ $14\frac{2}{5}$ fifth graders per class

   Ⓑ 18 fifth graders per class

   Ⓒ 24 fifth graders per class

   Ⓓ 216 fifth graders per class

21. Eduardo rides his bicycle for 6 hours. What was Eduardo's average speed if he rides a distance of 84 miles? Use the formula $d = r \times t$. **(pp. P267–P268)**

   Ⓐ 504 mi per hr

   Ⓑ 90 mi per hr

   Ⓒ 78 mi per hr

   Ⓓ 14 mi per hr

Understand Integers

Essential Question How can you use positive and negative numbers to represent real world quantities?

🔓 UNLOCK the Problem REAL WORLD

Connect You have used a number line to show 0 and whole numbers. You can extend the number line to the left of 0 to show the **opposites** of the whole numbers. For example, the opposite of ⁺3 is ⁻3. Any whole number or the opposite of a whole number is called an **integer**.

- How can you tell whether a number is an integer or not?

  _____

  _____

  _____

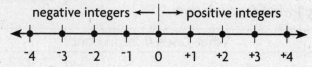

negative integers ←—|—→ positive integers

⁻4 ⁻3 ⁻2 ⁻1 0 +1 +2 +3 +4

Negative integers are written with a negative sign, ⁻. Positive integers are written with or without a positive sign, ⁺.

🔑 Example 1

The temperature in Fairbanks, Alaska, was 37 degrees below zero. Write an integer to represent the situation.

STEP 1 Decide whether the integer is positive or negative.

The word _____ tells me that the integer is _____.
_____

STEP 2 Write the integer: _____.

So, the temperature in Fairbanks was _____ degrees.

🔑 Example 2

The Koala Bears gained 11 yards on a football play. Write an integer to represent the situation. Then, tell what 0 represents in that situation.

STEP 1 Decide what positive integers and negative integers represent.

Positive integers represent yards _____.

Negative integers represent yards _____.
_____

STEP 2 Decide what 0 represents.

So, 0 means yards were neither _____

nor _____.

Math Talk **Identify** some words that might tell you that an integer is negative.

Share and Show

Write an integer to represent the situation.

1. a loss of $25

   The word *loss* represents an integer that is

   _____.

   The integer that represents the situation

   is _____.

2. 73 degrees above zero _____

3. 200 feet below sea level _____

4. a profit of $76 _____

Write an integer to represent the situation. Then, tell what 0 represents.

Situation	Integer	What Does 0 Represent?
5. The passenger jet flew at an altitude of 34,000 feet.		
6. Zack lost 45 points on his first turn.		
7. Craig was 20 minutes early for his appointment.		

On Your Own

Write an integer to represent the situation.

8. the temperature went up 2 degrees _____

9. 11 feet below sea level _____

10. an increase of 37 students _____

11. 15 seconds before rocket liftoff _____

Write an integer to represent the situation. Then, tell what 0 represents.

Situation	Integer	What Does 0 Represent?
12. Amelia earned $1,200 in one week.		
13. The coal was 2 miles below ground level.		
14. The alarm clock rang 5 minutes early.		

Problem Solving REAL WORLD

15. Gina withdrew $600 from her checking account to pay for her new guitar. What integer can you write to represent the withdrawal? What does 0 represent?

_____

Name _____

Write and Evaluate Expressions

Essential Question How can you write and evaluate expressions?

🔑 UNLOCK the Problem REAL WORLD

Montel hires Shea to buy some tools for him at the hardware store. Montel will pay Shea $5 more than the cost of the tools she buys.

A. How can you represent this payment as an expression?

B. How can you use the expression to calculate what Montel will pay Shea?

- The problem states that Montel will pay $5 *more than cost*. What operation do the words *more than* suggest?

_____

🔒 **Write an expression for what Montel will pay.**

STEP 1 Choose a variable and explain what it stands for.

Let *c* equal the cost of the tools.

_____

STEP 2 Write a word expression.

$5 more than the cost.

_____

STEP 3 Replace the word expression with an addition expression using *c*.

$5 + c$

So, an expression that tells how much Montel owes Shea is

$5 + c$ ___.

5 dollars more than the cost

$5 + c$

Try This! If the tools cost a total of $18, how much will Montel pay Shea?
Evaluate the expression $5 + c$ for $c = 18$.

STEP 1 Write the expression. _____

_____

STEP 2 Replace *c* with _____. $5 + $ _____

_____

STEP 3 Add to evaluate. $5 + 18 = $ _____

So, Montel will pay Shea _____.

Math Talk What key words might tell you that you need to use addition in a word problem?

Share and Show

Write an expression.

Tallahassee's temperature is 15 degrees less than the temperature in Miami.

1a. What operation does the phrase *less than* suggest?

_____

1b. Write a word expression:

_____

1c. Write an expression for Tallahassee's temperature. Let *m* stand for the temperature in Miami.

_____

1d. Evaluate the expression for Tallahassee's temperature for $m = 90$.

_____

Evaluate each expression for the value given.

2. $b - 45$ for $b = 70$

_____

3. $13 + a$ for $a = 40$

_____

On Your Own

Write an expression.

4. Zeke has some tropical fish, *f*. Dean gave Zeke 5 new fish. How many fish does Zeke have now?

_____

5. Myra had some candles, *c*. She used up 12 of them. How many candles does Myra have now?

_____

Evaluate each expression for the value given.

6. $s - 18$ for $s = 80$

_____

7. $49 + k$ for $k = 31$

_____

8. $w \times 6$ for $w = 13$

_____

9. $60 \div n$ for $n = 20$

_____

10. $t \times 12$ for $t = 8$

_____

11. $r - 25$ for $r = 110$

_____

Problem Solving REAL WORLD

12. Keith is 2 inches shorter than his sister. If *s* represents his sister's height, what expression can you write that represents Keith's height?

_____

Name _____

Understand Inequalities

Essential Question How can you use inequalities to solve problems?

 UNLOCK the Problem REAL WORLD

Every morning, Bobbi's Hot Bagels makes a special claim. All bagels Bobbi's sells will be warm and less than 9 minutes old. What **inequality** can you write to represent in whole minutes how old Bobbi's bagels are?

An inequality is a number sentence that compares two unequal quantities and uses the symbols $<$, $>$, $\leq$, or $\geq$.

- What clue words tell you that this problem involves an inequality?

  _____

🔑 **Write an inequality using a variable.**

STEP 1 Write the inequality in words.   time ⟶ is less than ⟶ 9

STEP 2 Replace *time* with the variable *t*. t ⟶ less than ⟶ 9

STEP 3 Replace the words *less than* with a *less than* ($<$) symbol. $t < 9$

Try This! Graph the solutions on the number line. Of 3, 6, 9, and 12, which numbers are solutions for $t < 9$?

STEP 1 In $t < 9$, replace t with 3. $t < 9$
Repeat the process for $t = 6, 9, 12$. $3 < 9$ ⟵ true

STEP 2 Identify the values that make $t < 9$ true. $6 < 9$ ⟵ true
True values are solutions: $t = 3, 6$. $9 < 9$ ⟵ false
False values are not solutions: $t \neq 9, 12$. $12 < 9$ ⟵ false

STEP 3 Graph the solutions on a number line.
Graph true values with filled circles.

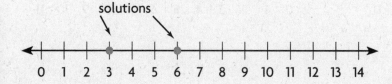

solutions

Math Talk How does the answer for the problem change if the inequality is "t is less than or equal to 9"?

Share and Show

Of 2, 5, and 8, which numbers are solutions for the inequality $x \geq 5$?
Graph the solutions on the number line.

1a. Replace x with 2. True or false?

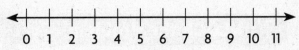

_____

1b. Replace x with 5. True or false?

_____

1c. Replace x with 8. True or false?

_____

Show two solutions for the inequality on a number line.

2. $a < 6$

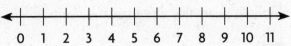

_____

On Your Own

Of 7, 10, and 13, which numbers are solutions for the inequality?

3. $m > 8$

_____

4. $b \leq 10$

_____

5. $c < 15$

_____

Of 0, 4, 6, and 11, which numbers are solutions for the inequality?

6. $d \geq 8$

_____

7. $r < 1$

_____

8. $s > 4$

_____

Show two solutions for the inequality on a number line.

9. $n \leq 6$

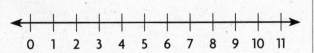

10. $x > 2$

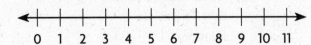

Problem Solving REAL WORLD

11. For her birthday party, Dina wants to invite at least 8 guests but not
more than 12 guests. How many guests might she have? Name all of
the possibilities.

_____

Name _____

✓ Checkpoint

Concepts and Skills

Write an integer to represent the situation. (pp. P271–P272)

1. a shark 125 feet below sea level _____
2. a bank deposit of 300 dollars _____

Write an integer to represent the situation. Then, tell what 0 represents. (pp. P271–P272)

Situation	Integer	What Does 0 Represent?
3. a gain of 13 yards by a football team	_____	
4. a temperature of 25 degrees below zero	_____	

Write an expression. Then evaluate the expression for the value given. (pp. P273–P274)

5. Miki has n dollars. Dora has 3 more dollars than Miki. How many dollars does Dora have? Evaluate for $n = 14$.

_____

6. Chip has s shells. Gina has 4 times as many shells as Chip. How many shells does Gina have? Evaluate for $s = 6$.

_____

Of 1, 3, 4, and 8, which numbers are solutions for the inequality? (pp. P275–P276)

7. $a < 7$ 8. $b \geq 3$ 9. $c > 4$ 10. $d \leq 8$

_____   _____   _____   _____

Problem Solving REAL WORLD

Filters are set up to sort pennies, dimes, and nickels. A penny is 19 mm wide, a dime is 17.9 mm wide, and a nickel is 21 mm wide. Coins less than 20 mm wide will pass through the first level, and coins less than 18.5 mm wide will pass through the second level. (pp. P275–P276)

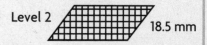

drop coins

Level 1 — 20 mm

Level 2 — 18.5 mm

11. If you drop a large number of all 3 coins from above, which coins will be caught at Level 1? Which coins will pass through?

_____

12. Which coins will be caught at Level 2? Which coins will pass through? _____

Fill in the bubble completely to show your answer.

13. The lowest temperature ever recorded in North Dakota was 60 degrees below zero Fahrenheit. Which integer represents the temperature? (pp. P271–P272)

   (A) 0

   (B) 60

   (C) ⁻60

   (D) ⁻0

14. In football, a team receives 3 points for each field goal it makes. Which expression shows the number of points a team will receive for making *f* field goals? (pp. P273–P274)

   (A) $3 + f$

   (B) $3 \times f$

   (C) $f - 3$

   (D) $f \div 3$

15. The elevation of Central City is 84 feet above sea level. Which integer is the opposite of 84? (pp. P271–P272)

   (A) 48

   (B) ⁺84

   (C) ⁻48

   (D) ⁻84

16. Uncle Louie is at least 1 inch shorter than Miriam, and at least 2 inches taller than Jeffrey. Jeffrey's height is 64 inches. Miriam is not more than 5 inches taller than than Jeffrey. Which answer choice could be Uncle Louie's height? (pp. P275–P276)

   (A) 65 inches

   (B) 67 inches

   (C) 69 inches

   (D) 70 inches

Name _____

Polygons on a Coordinate Grid

Essential Question How can you plot polygons on a coordinate grid?

Connect You have learned to plot points on a coordinate grid. You can use that skill to plot polygons on a coordinate grid.

 UNLOCK the Problem REAL WORLD

Camille is designing an indoor greenhouse on a coordinate grid. The floor of the greenhouse is a polygon. The vertices of the polygon can be graphed using the coordinates shown in the table. Plot and describe the floor of the greenhouse.

x	y
10	1
2	6
2	1
6	10
10	6

- What do x and y represent in the table?

_____

_____

_____

🔑 **Plot the polygon on a coordinate grid.**

STEP 1 Write ordered pairs.

Use each row of the table to write an ordered pair.

(10, 1), (2, _____), (_____ , _____),

(_____ , _____), (_____ , _____).

_____

STEP 2 Graph a point for each pair on the coordinate grid.

_____

STEP 3 Connect the points.

So, the floor of the greenhouse is a _____.

- **What if** the greenhouse floor had only four of the five vertices given in the table and did not include (6, 10). What would the shape of the floor be? _____

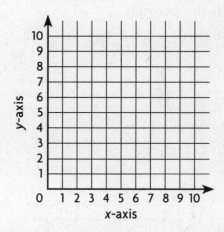

- A parallelogram on a coordinate grid has vertices at (3, 4), (6, 1), and (8, 4). What are the coordinates of the fourth vertex? **Explain** how you found the answer.

_____

_____

_____

_____

_____

Math Talk **Suppose** you know the vertices of a polygon. How can you identify what type of polygon it is without plotting the vertices on a coordinate grid?

Share and Show

Plot the polygon with the given vertices on a coordinate grid.
Identify the polygon.

1. (9, 6), (1, 7), (3, 1)

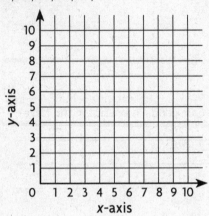

_____

2. (1, 6), (8, 4), (1, 4), (8, 6)

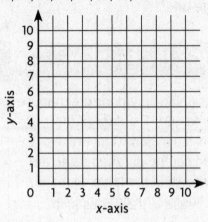

_____

On Your Own

Plot the polygon with the given vertices on a coordinate grid.
Identify the polygon.

3. (2, 10), (10, 2), (10, 10), (2, 2)

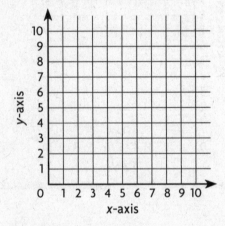

_____

4. (10, 4), (2, 10), (3, 1), (8, 0), (7, 10), (1, 7)

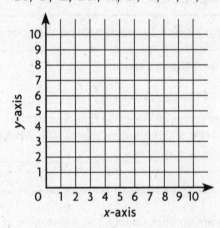

_____

Problem Solving REAL WORLD

5. A football field is a rectangle measuring 300 ft by 160 ft. Each unit on
a coordinate grid represents 1 foot. (0, 0) and (0, 160) are two of the
coordinates of a football field drawn on the grid. What are the coordinates
of the other two vertices?

_____

Name _____

Area of a Parallelogram

Essential Question How can you find the area of a parallelogram?

Connect You have learned that the area of a rectangle with base *b* and height *h* is $A = b \times h$. The rectangle shown has a base of 5 units and a height of 3 units. So, its area is $A = 5 \times 3 = 15$ square units. You can use what you have learned about the area of a rectangle to find the area of a parallelogram.

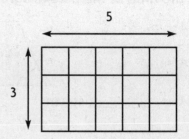

UNLOCK the Problem REAL WORLD

The souvenir stand at Mighty Grasshopper basketball games sells parallelogram-shaped pennants. Each pennant has a base of 12 inches and a height of 5 inches.

🔑 Activity Find the area of the parallelogram.

Materials ■ grid paper ■ scissors

STEP 1 Draw the parallelogram on grid paper and cut it out.

STEP 2 Cut along the dashed line to remove a right triangle.

STEP 3 Move the right triangle to the right side of the parallelogram to form a rectangle.

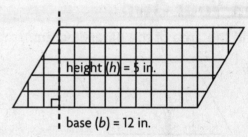

STEP 4 The base of the rectangle measures _____ inches.

The height of the rectangle measures _____ inches.

The area of the rectangle is

$12 \times$ _____ = _____ square inches.

• **Explain** why the area of the parallelogram must equal the area of the rectangle.

_____

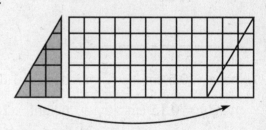

So, the area of a pennant is

_____ × _____ = _____ square inches.

Math Talk **Explain** how to find the area of a parallelogram if you know the base and the height of the figure.

Share and Show

Find the area of the parallelogram.

1. $A = b \times h$

$A = 8 \times 4$

$A = $ _____ sq cm

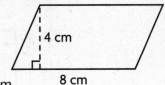

4 cm

8 cm

2.

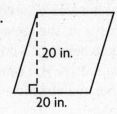

20 in.

20 in.

$A = $ _____ sq in.

3.

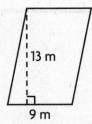

13 m

9 m

$A = $ _____ sq m

4.

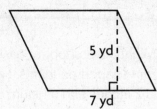

5 yd

7 yd

$A = $ _____ sq yd

On Your Own

Find the area of the parallelogram.

5.

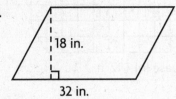

18 in.

32 in.

$A = $ _____ sq in.

6.

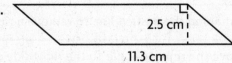

2.5 cm

11.3 cm

$A = $ _____ sq cm

7. base = 0.6 cm

height = 0.15 cm

$A = $ _____ sq cm

8. base = 1.8 m

height = 2.9 m

$A = $ _____ sq m

9. base = $\frac{1}{2}$ ft

height = $\frac{3}{8}$ ft

$A = $ _____ sq ft

10. base = $4\frac{1}{4}$ in.

height = 20 in.

$A = $ _____ sq in.

Problem Solving REAL WORLD

11. Carla made a border for her garden using parallelogram-shaped tiles. Each piece had a base of 4 in. and a height of $2\frac{1}{2}$ in. She used 85 tiles. What was the total area of the border?

_____

Name _____

Median and Mode

Essential Question How can you describe a set of data using median and mode?

The **median** of a set of data is the middle value when the data are written in order. For example, a baseball team scored 6, 2, 6, 0, and 3 runs in five games. The median is 3 runs: 0, 2, ③, 6, 6.

If there is an even number of data items, the median is the sum of the two middle items divided by 2.

The **mode** of a data set is the data value or values that occur most often. A data set may have no mode, one mode, or several modes. The mode of the data set of baseball runs is 6.

🔓 UNLOCK the Problem REAL WORLD

For the Science Fair, Ronni grew 9 sweet pea plants under different conditions. Here are the plants' heights, in centimeters: 11, 13, 6, 9, 15, 7, 9, 17, 12.

What are the median and mode of the data?

- How can you find the median if there is an even number of data items?

  _____

 Find the median and mode.

STEP 1 Order the heights from least to greatest.

6, 7, _____, _____, _____, _____, _____, _____, _____
_____

STEP 2 Circle the middle value.

So, the median is _____ centimeters.
_____

STEP 3 Identify the data value that occurs most often. _____ occurs two times.

So, the mode is _____ centimeters.

Math Talk Give an example of a data set with two modes.

Try This! Find the median and mode of the numbers: 8, 11, 13, 6, 4, 3.

STEP 1 Order the numbers from least to greatest.

_____, _____, _____, _____, _____, 13
_____

STEP 2 There is an even number of data items, so divide the sum of the two middle items by 2. $\frac{6 + \_\_}{2} = \frac{\_\_}{2} = \_\_\_\_$

So, the median is = _____.
_____

STEP 3 _____ data value appears more than once.

So, the data set has _____ mode.

Share and Show

Find the median and the mode of the data.

1. puppies' weights (pounds): 8, 3, 5, 3, 2, 6, 3

   Order the weights: _____

   The median, or middle value, is _____ pounds

   The mode, or most common value,

   is _____ pounds.

2. numbers of students in math classes:
   25, 21, 22, 18, 23, 24, 25

   median: _____ students

   mode: _____ students

3. numbers of 3-point baskets made:
   2, 0, 5, 4, 5, 2, 5, 2

   median: _____ 3-point baskets

   mode: _____ 3-point baskets

4. movie ticket prices ($):
   8, 8, 6, 8, 7, 6, 8, 10, 8, 6

   median: $_____

   mode: $_____

On Your Own

Find the median and the mode of the data.

5. ages of first 10 U.S. presidents
   when inaugurated:
   57, 61, 57, 57, 58, 57, 61, 54, 68, 51

   median: _____ years

   mode: _____ years

6. weights of rock samples (pounds):
   39, 28, 21, 47, 40, 33

   median: _____ pounds

   mode: _____ pounds

7. lengths of humpback whale songs (minutes):
   25, 29, 31, 22, 33, 31, 26, 22

   median: _____ minutes

   mode: _____ minutes

8. Sascha's test scores:
   90, 88, 79, 97, 100, 97, 92, 88, 85, 92

   median: _____

   mode: _____

Problem Solving

9. Adrian recorded the daily high temperatures the first two weeks of July.
   What were the median and mode of her data?

   median: _____ °F

   mode: _____ °F

Daily High Temperatures (°F)						
101	99	98	96	102	101	98
101	98	95	100	102	98	102

Name _____

Finding the Average

Essential Question How can you find the average of a set of values?

An average of a set of data can be found by finding the sum of the group of numbers from the data and then dividing by the number of addends.

For example, if Anne scores 21 points, 22 points, and 17 points in 3 different basketball games, she scores an average of 20 points per game. This is because $21 + 22 + 17 = 60$, and $60 \div 3$, the total number of points divided by the number of games, is 20.

🔓 UNLOCK the Problem   REAL WORLD

Jonathon and Pilar are practicing to be a juggling team. The table shows the number of seconds they were able to keep 4 balls in the air without making a mistake. What was the average number of seconds they were able to juggle?

Trial	Seconds
a	32
b	8
c	62
d	55
e	13

- How many trials did they record?

  _____

🔑 **Find the average of the times.**

STEP 1   Find the sum of the seconds.

$32 + 8 + 62 + 55 + 13 = 170$

STEP 2   How many numbers did you add?

5 numbers

STEP 3   Divide the sum by the number of addends. $5)\overline{170}$ = 34

So, the average time that Jonathon and Pilar kept 4 balls in the air was __34__ seconds per trial.

Try This! Find the average of 61, 99, 106, 3, 44, and 89.

STEP 1   Find the sum.

$61 + 99 + 106 + 3 + 44 + 89 = $ _____

STEP 2   Divide the sum by the number of addends.

$402 \div 6 = $ _____

So, the average of 61, 99, 106, 3, 44, and 89 is _____.

Math Talk Use the jugglers' average time per trial. What might you expect of them in their next trial?

Share and Show

Tommy's basketball scoring record is shown for this month. What was the average number of points that Tommy scored per game?

1a. Find the sum of the points Tommy scored.

_____

Game	1	2	3	4	5	6	7	8
Points	24	11	31	14	9	21	18	8

1b. How many numbers did you add to find the sum in Exercise 1?

_____

1c. Divide the sum by the number of games. What is the average number of points per game?

_____

Find the average of the set of numbers.

2. 6, 9, 14, 4, 12

_____

3. 44, 55, 33, 22, 40, 40

_____

On Your Own . . . .

Find the average of the set of numbers.

4. 4, 8, 12, 14, 15, 19

_____

5. 28, 20, 31, 17

_____

6. 100, 140, 60, 120, 180

_____

7. 17, 91, 49, 73, 115, 27

_____

8. 5, 8, 13, 4, 22, 6, 0, 5, 9

_____

9. 637, 492, 88, 743

_____

10. 2,439; 801; 1,508; 0

_____

11. 13, 12, 11, 13, 15, 13, 19, 22, 13, 19

_____

12. 78, 61, 51, 99, 8, 112, 76, 32, 59

_____

13. Find the average temperature.

_____

Day	1	2	3	4	5	6	7
Temperature (°F)	48	59	38	53	61	61	44

Problem Solving

14. In the temperature table above, suppose the temperature for the next 2 days was 70 degrees. By how much would this change the average temperature over the entire period?

_____

Name _____

Histograms

Essential Question How can you use a histogram to organize data?

🔑 Activity The table below shows the ages of the members of a bicycle club. Make a **histogram** of the data. A histogram is a bar graph that shows how often data occur in intervals.

> **Math Idea**
> In a histogram, the bars touch because they represent continuous intervals.

Ages of Members in a Bicycle Club

34	38	29	41	40	35	50	20	47	22	19	21	18	17
26	30	41	43	52	45	28	25	39	24	23	25	50	59

STEP 1 Make a frequency table with intervals of 10. Fill in the frequencies.

STEP 2 Choose an appropriate scale and interval for the vertical axis, and list the intervals on the horizontal axis. Label each axis.

STEP 3 Draw a bar for each interval. Give the histogram a title.

Ages	Tally	Frequency
10–19	|||	
20–29	⊞⊞ ||||	
30–39	⊞||	
40–49	⊞||	
50–59	||||	

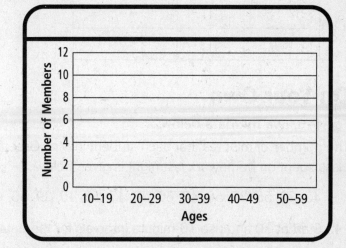

- **What if** you changed the histogram to show four age groups with 12-year intervals?

  How would the histogram change?

  _____

  _____

  _____

> **Math Talk** **Explain** how a histogram and a bar graph with categories are different.

Share and Show

For 1–3, use the data below.

The number of vacation days that each employee of a
company took last summer is given below.

2, 5, 6, 11, 3, 5, 7, 8, 10, 1, 4, 6, 10, 5, 12, 15, 6, 8, 7, 14

1. Start at 1 day and use 4 days for each interval. List the intervals.

_____

2. Complete the frequency table.

Number of Days	Tally	Frequency
1–4	||||	
5–8	~~||||~~ ~~||||~~	
9–12	||||	
13–16	||	

3. Complete the histogram.

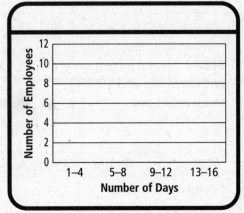

On Your Own

For 4–6, use the data below.

The number of minutes that each student in Mrs. Green's
class spent on homework last night is given below.

45, 30, 55, 35, 50, 48, 60, 38, 47, 56, 40, 39, 55, 65, 49, 34, 35

4. Start at 30 and use 10-minute intervals for the data. List the intervals.

_____

5. Make a frequency table of the data.

6. Make a histogram of the data.

Problem Solving REAL WORLD

7. The number of words per minute that one class of students
   typed is given below.

   30, 45, 28, 35, 48, 37, 41, 44, 34, 29, 25, 32, 40, 45, 39, 49

   What are reasonable intervals for the data?

_____

Name _____

Analyze Histograms

Essential Question How can you analyze data in a histogram?

🔓 UNLOCK the Problem REAL WORLD

The histogram shows the number of items sold at a garage sale within each price range.

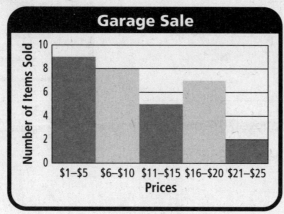

! ERROR Alert

Remember to read the intervals. For some questions, you may need to combine data from two or more intervals in order to answer the question.

🔑 **How many of the items sold cost $6 to $10?**

• Find the interval labeled $6–$10.

• Find the frequency.

• The bar for $6–$10 shows that _____ items were sold.

So, _____ of the items sold cost $6 to $10.

🔑 **How many of the items sold cost $16 to $25?**

• Find the frequencies for the intervals labeled $16–$20 and $21–$25.

• The bar for $16–$20 shows that _____ items were sold. The bar for $21–$25

  shows that _____ items were sold.

• Add the frequencies.

  7 + _____ = _____

So, _____ of the items sold cost $16 to $25.

Math Talk **Explain** why you cannot tell from the histogram the total amount of money that was made during the garage sale.

Share and Show

For 1–3, use the histogram at the right.

1. The histogram shows the number of days in one month whose temperatures were within each temperature range. On how many days was the temperature at or above 70°F?

   - List the bars that represent temperatures at or above 70°F.

     _____ and _____

   - The frequency for interval 70–74 is _____, and the

     frequency for interval 75–79 is _____.

   - Add the frequencies. _____ + _____ = _____

The daily high temperature was at or above 70°F on _____ days.

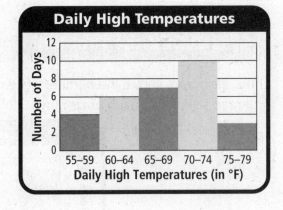

2. On how many days was the temperature 65°F to 69°F?

   _____

3. On how many days was the temperature less than 65°F?

   _____

On Your Own ·

For 4–5, use the histogram at the right.

4. Which interval has the greatest frequency? _____

5. How many days did Maxine ride the stationary bike for 30

   or more minutes? _____

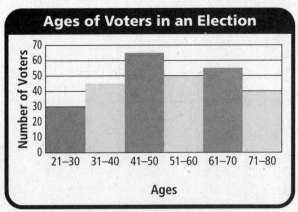

Problem Solving

For 6–7, use the histogram at the right.

6. How many people voted in the election?

   _____

7. How many more voters were there from ages
   41–50 than from ages 21–30?

   _____

Name _____

✓ Checkpoint

Concepts and Skills

1. Plot and identify the polygon with vertices at (4, 0), (8, 7), (4, 7), and (8, 0). **(pp. P279–P280)**

   _____

2. A parallelogram has a base of 8.5 cm and a height of 6 cm. What is the area of the parallelogram? **(pp. P281–P282)**

   _____

3. Find the median and mode of Erin's math sores: 93, 88, 85, 93, 100, 95, 85, 89. **(pp. P283–P284)**

   median _____ mode _____

4. Find the average of the following temperatures: 59°F, 66°F, 59°F, 67°F, 54°F, 64°F, 72°F. **(pp. P285–P286)**

   average _____

For 5–7, use the data below. (pp. P287–P290)

The math test scores for Miss Jackson's class are given below.

88, 94, 86, 78, 65, 83, 71, 74, 92, 73,
95, 71, 100, 98, 68, 85, 81, 93, 89, 84

5. Make a histogram for the data using intervals of 10.

6. Which interval has the greatest frequency?

   _____

7. How many students received grades greater than 80? _____

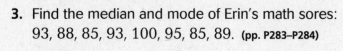

Problem Solving REAL WORLD

For 8–9, use the histogram. The histogram shows the times that people wake up in the morning. (pp. P287–P290)

8. How many people were surveyed? _____

9. How many more people surveyed wake up between 6:30 and 6:59 than between 7:30 and 7:59?

   _____

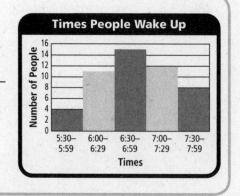

Times People Wake Up

© Houghton Mifflin Harcourt Publishing Company

Fill in the bubble completely to show your answer.

10. On a map of the town of Barton, City Hall Park has three of its four
    vertices at (15, 0), (5, 0), and (15, 9). City Hall Park is a rectangle.
    What are the coordinates of the park's fourth vertex? (pp. P279–P280)

    Ⓐ (5, 9)

    Ⓑ (9, 5)

    Ⓒ (5, 15)

    Ⓓ (9, 15)

11. A window at an art gallery is shaped like a parallelogram. The base
    measures 1.2 meters and the height measures 0.8 meters. What is
    the area of the window? (pp. P281–P282)

    Ⓐ 0.48 sq m

    Ⓑ 0.96 sq m

    Ⓒ 1.92 sq m

    Ⓓ 2.0 sq m

12. The ages of the members of the Chess Club are given below. What is
    the median age? (pp. P283–P284)
    13, 9, 10, 9, 14, 13, 8, 9

    Ⓐ 9

    Ⓑ 9.5

    Ⓒ 10

    Ⓓ 10.5

13. The histogram shows the ages of runners in
    a half-marathon. How many runners are
    between the ages of 21 and 40? (pp. P289–P290)

    Ⓐ 24

    Ⓑ 30

    Ⓒ 42

    Ⓓ 54

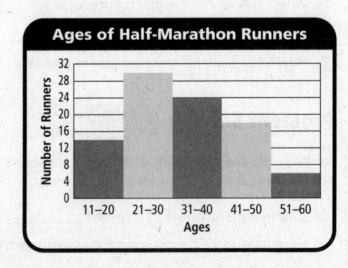